Logic of Science in Psychoanalysis

LOGIC of SCIENCE in PSYCHOANALYSIS

BENJAMIN B. WOLMAN

New York COLUMBIA UNIVERSITY PRESS *1984*

Library of Congress Cataloging in Publication Data

Wolman, Benjamin B.
Logic of science in psychoanalysis.

Bibliography: p.
Includes indexes.
1. Psychoanalysis—History. 2. Personality—History.
3. Child psychology—History. I. Title.
BF175.W64 1984 150.19′5 84-1891
ISBN 0-231-05744-X (alk. paper)

Columbia University Press
New York Guildford, Surrey

Printed in the United States of America

Clothbound editions of Columbia University Press Books are
Smyth-sewn and printed on permanent and durable acid-free paper

CONTENTS

Preface vii

Part I. Philosophical and Historical Foundations

ONE. Methodological and Philosophical Considerations 3

TWO. Cultural-Historical Perspectives 40

Part II. Freud's Theory

THREE. First Principles 61

FOUR. Mental Strata 77

FIVE. The Driving Forces 101

SIX. Developmental Stages 115

SEVEN. Personality Structure 136

EIGHT. Social and Cultural Issues 160

Part III. Freudian and Neo-Freudian Theories

NINE. Hartmann's Ego Psychology 173

TEN. Klein's Developmental Theory 192

ELEVEN. Erikson's Quest for Identity 206

TWELVE. Mahler's Separation and Individuation Theory 214

THIRTEEN. Wolman's Interactional Theory 228

Part IV. Non-Freudian Theories

FOURTEEN. Adler's Individual Psychology 251

FIFTEEN. Jung's Analytic Psychology 266
SIXTEEN. Horney's "New Ways" in Psychoanalysis 299
SEVENTEEN. Sullivan's Interpersonal Theory 309
EIGHTEEN. Quo Vadis, Psychoanalysis? 323
Name Index 327
Subject Index 331

PREFACE

Freud's mission was a relentless search for truth. His intellectual heritage can be summarized in two words: "Sapere aude!" (Have the courage to know). Some scientists avoid the complex irrationality of human nature; some shy away from the depth of unconscious motivation; some fear the volcanos of human emotions; but nothing could stop Freud.

Freud was an iconoclast. Freud crossed swords and words with psychology, psychiatry, anthropology, biology, sociology, politics, theology, art, and literature. He developed a complex and far reaching cross-disciplinary system that inspired many and alienated many.

Freud laid the foundation for several new and different theories—some by his followers, some by dissidents. Heinz Hartmann and Franz Alexander, Margaret Mahler and Heinz Kohut, and many others have continued and further developed Freud's conceptual system. Alfred Adler and Carl Gustav Jung, Harry Stack Sullivan and Karen Horney, and many others have introduced totally different theories. It is impossible to do justice to the great number and ever-expanding branches and briars of the psychoanalytic thought, and the present book is limited to a few representative theoretical systems.

Certainly the theory Freud built over decades requires a fresh look. Several scholars have embarked on a search for new empirical data and new interpretations of data. The problems of anxiety and narcissism, subliminal perception and biological drives, sleep and dreams, psychosomatic medicine and neurological determinants of behavior, and the entire field of psychopathology have been critically reexamined by J. Arlow, J. Brenner, M. Gill, R. R. Holt, G. S. Klein, H. Kohut, L. S. Kubie, R. J. Langs, M. Mayman, K. H. Pribram, I. Sarnoff, R. Schafer, and many others who introduced fresh elements some in agreement with Freud's basic tenets, some opening new vistas in the study of human nature. There is an ever-growing body of experimental research and logical analysis of psychoanalysis and refreshing efforts to bring it up to date.

Scores of scholarly papers and monographs bear witness to the vitality of psychoanalytic thoughts.

The task of the present book was stated in the first sentence of the preface: the search for truth. The logic and the scientific method in psychoanalysis and related theoretical systems is our subject matter. The aim of this book is not to suggest how psychoanalytic thought could or should be modified, but to describe how the principles of formal logic and philosophy of science were applied in psychoanalytic theories. The conceptual systems and the interpretation of available data by Freud and some of his disciples and opponents are the subject matter of this book. Hence its title: *Logic of Science in Psychoanalysis.*

The entire book is divided into four parts. The first part deals with philosophical and cultural-historical perspectives; the second describes Freud's psychological theory; the third describes theories rooted in Freud's principles; the fourth delves into non-Freudian theories. The last chapter describes a few representative issues that shed light on the historical and cultural perspectives of psychoanalysis.

[PART I]
PHILOSOPHICAL AND HISTORICAL FOUNDATIONS

[ONE]

Methodological and Philosophical Considerations

THE history of science is full of anecdotic stories, such as Archimedes discovering that his body weighed less in a bathtub than on the surface of the earth and Newton's apple leading to the discovery of the laws of gravitation. Even highly sophisticated researchers did not necessarily follow the rules of logic and methodology of science. Consider Pavlov's and Thorndike's studies, which were conducted without experimental controls; or Freud's discoveries derived from listening to a totally coincidental sample of mental patients and from scattered efforts of interpreting his own dreams; or Piaget's observations and investigations of children, which did not apply the rules of the experimental procedures. A great many important scientific discoveries, such as Darwin's theory of evolution and Galileo's and Copernicus' discoveries, did not evolve out of systematic and precise observation nor were they a result of planned experimentation. Apparently "unsupported speculation" brought often spectacular results (Feyerabend 1963:5).

The history of psychology resembles the history of other sciences only to a certain extent. The history of science is a history of how certain human beings went about the discovery of scientific truth in various branches of science, how they observed, reasoned, abstracted, tried and erred. But observation, reasoning, abstraction, and all other activities that scientists perform belong to psychology. All of them are part of human behavior, specifically cognitive behavior, and the history of psychology is the history of what certain scientists had discovered about themselves and others.

The history of psychology is even more involved, for logic and scientific method are inseparably interlocked with the psychology of cognitive processes. Whenever scientists had tried to establish objective rules for their conduct in laboratories, observatories, study rooms, and lecture halls, they were both the lawmakers who had set these rules and the subject of these rules.

There is no reason to ascribe superhuman, divine, and infallible intellectual powers to a Harvey, Newton, Boyle, Darwin, Mendel, Freud, Pavlov, and Einstein; all were members of the *Homo sapiens* species, driven by curiosity and ambition, vanity and altruism, the need to earn a living and the fear of want, self-confidence and inferiority feeling, anxiety, and many other human motives. Scientists do not live in a vacuum. All scientists were brought up within a certain cultural system, were taught by certain teachers, indoctrinated by certain cultural values, and faced with problems transmitted by their ancestors and contemporaries.

Take, for instance, Descartes. Descartes was concerned with the separation of physics from theology, for this was the burning problem at the times of religious wars and beginnings of scientific inquiry. Whether Descartes was truly religious or was afraid to express his doubts in religious matters is still a controversial issue. Descartes' times were neither conducive to an outright rejection of religious dogmas nor to a naive theological interpretation of the universe. One hundred years later the Encyclopedists embarked upon a secular interpretation of the universe without resorting to religious concepts. But one hundred years before Descartes no one could have predicted what Descartes would learn from Harvey on the nature of living organisms.

A similar reasoning applies to psychoanalysis, as it will be explained later. Freud was influenced by Darwin; Adler by Vaihinger; Jung by oriental philosophy. All of them received their education in universities of central Europe and all of them were exposed to philosophical problems of their times.

Rationalism and Empiricism

Scientists do not possess a monopoly on truth, and truthful statements are made daily by a great many sources, such as the radio which tells the time and quotes the stock market, the weatherman who reports yesterday's weather, and the telephone directory.

Scientists, with some unfortunate exceptions, do not study the obvious nor do they discover what is common knowledge. Scientists seek the hitherto unknown information concerning objects and events (that is, what is going on with objects). The term "objects" includes living organisms, and the term "events" includes behavior of organisms. This definition does not necessarily contradict Wilhelm Stern's (1924) distinction between *Person* (person) and *Sache* (thing); it merely stresses the fact that psychology is part and parcel of the natural sciences and it studies a certain fraction of what is going on in the universe (Wolman 1981: ch. 15).

The fact that several important scientific discoveries were made accidentally does not indicate that scientists must rely on "inspiration" in order to produce information. The specific trait of science as a part of cognitive behavior is the *search for evidence.* The search for proof for whatever information was obtained distinguishes science from all other human activities such as beliefs, intuition, prophesies, and plain gossip.

The aim of scientific inquiry, usually called *research,* is the search for knowledge and for proof that the given information was correct. Traditionally there have been two approaches to the problem of acquisition of knowledge, namely *speculative rationalism* and *sensualistic empiricism.*

The point of view of rationalism was represented by Descartes and Spinoza. According to Descartes: "Our inquiries should be directed not to what others have thought, nor to what we ourselves conjecture, but to what we can clearly and perspicuously behold and with certainty deduce; for knowledge is not won in any other way" (Rule III). According to Descartes, only mathematics succeeds in producing "evident and certain" reasons. Descartes believed that "the power of forcing a good judgment and of distinguishing the true from the false . . . is by nature equal in all men" (p. 81). Hence Descartes logically deduced his own existence (Cogito ergo sum) and the existence of God and universe without referring to empirical evidence.

Galileo's mathematics influenced Spinoza, who developed a system of definitions, axioms, and propositions not related to empirical data. Spinoza's philosophy was rationalistic, his evidence speculative, and his method of reasoning mathematico-deductive. Spinoza believed that his first premises are self-evident; he called them "adequate ideas" which need no proof (Proposition VII).

Locke was critical of Descartes and rationalists. He rejected the innate ideas and ridiculed "remote speculative principles" that are "like curious imagery men sometime see in the clouds" (Fox Bourne 1876 1:224). "All ideas come from sensation or reflection," wrote Locke. "Our observation employed either about external sensible objects or about the internal operations of our minds perceived and reflected on by ourselves, is that which supplies our understanding with all the materials of thinking. These are the two fountains of knowledge" (2[2]:1). Our sensory apparatus is the sole source of information: Nihil est in intellectu quod in sensibus non fuerat (There is nothing in our intellect that was not in our senses), Locke wrote.

Hume went further than Locke in the emphasis on sensory origin of knowledge. Hume distinguished between impressions and sensations, such as hearing, seeing, etc., and ideas. Ideas are derived from impressions. Hume pointed to the fact that since "external objects as they appear to the senses, give us no idea of power or necessary connection in particular instances, let us see, whether this idea be derived from reflection on the operations of our own minds and be copied from any internal impressions" (3[2]:1).

Kantianism

One may see in Spinoza a radical version of Descartes' and in Hume a radical version of Locke's philosophy; certainly empiricism and rationalism do not blend. Kant, however, fused the two systems into one. The rationalists believed that knowledge is a system of *a priori* given propositions; the empiricists sought *a posteriori* observations. Kant's *Critique of Pure Reason* rejected both points of view and offered a solution that weighed heavily upon any future analysis of the problem.

According to Kant:

> Though all our knowledge begins with experience, it does not follow that it all arises out of experience. For it may well be that even our empirical knowledge is made up of what we receive through impressions and of what our own faculty of knowledge (sensible impressions serving merely as the occasion) supplies from itself. If our faculty of knowledge makes any such addition, it may be that we are not in position to distinguish it from the raw material. . . . This, then, is a question which at best calls for closer examination, and does not allow of any offhand answer:—whether there is any knowledge that is thus independent of experience and even of all impressions of the senses. Such knowledge is entitled *a priori,* and distinguished from the empirical, which has its sources *a posteriori,* that is, in experience. (p. 1)

Kant accepted the empiricists' view that sensations are the source of knowledge. He believed, however, that this knowledge based on human senses is a knowledge of phenomena perceived in a manner determined by the nature of the human mind. The human mind perceives things in terms of time and space and the categories of relation, quantity, quality, and modality.

The mind acquires knowledge in terms which are given a priori and independent of any experience. The only true science is, therefore, mathematics, for its propositions are nonempirical, synthetic, and a priori. The world of objects-in-themselves is inaccesible to empirical sciences; what the empirical sciences study are not the things as they are but mere appearances, *phenomena* perceived in terms of the transcendental mind.

Speculative philosophy reached its peak with Kant's distinction between *noumena,* that is, nature-as-it-is (*Ding an sich,* usually translated "thing-in-itself"), and *phenomena,* that is, nature as perceived. According to Kant, laws of nature were not a part of nature but abstract terms of time, space, quantity, causation, etc., embedded in the *Absoluter Geist* and prescribed to the universe (Cassirer 1923; Wolman 1968).

Most philosophers of the nineteenth century, especially in German-speaking countries, were Kantians and neo-Kantians. Hegel dictated dialectic logic to the universe and ascribed dialectic laws to human history assuming that whatever happens, happens in a trilogy of thesis, antithesis, and synthesis. When Hegel's highly biased and distorted historical presentation was criticized as full of factual errors, and his theory blamed for distortion of reality, Hegel's famous reply was "Desto schlimmer für die Wirklichkeit" (It's too bad for reality).

Schopenhauer's system has shown even less consideration for the real world. Schopenhauer pictured the world as an act of will and representation *(Die Welt als Wille und Vorstellung),* reducing reality to a product of one's wishes and ideas. The universe is a mere representation, an idea created by perceiving mind; only I, my will, is a thing-in-itself. "The will is the thing-in-itself, the inner content, the essence of the world," Schopenhauer wrote. "Life, the visible world, the phenomena, is only the mirror of the will" (Book 4, section 54).

Phenomenology

Kant put the perceiving transcendental mind into the center of the universe. According to Kant's anti-Copernican revolution, the world revolves around the perceiving "absolute" mind that imposes on nature its categories of time and space, causality and quality. Empirical sciences were reduced to a study of mere appearances, and mathematics was to become the only valid and objective science.

Those philosophers and scientists who wished to follow the road of empirical science but could not break the spell of Kant's heritage sought various ways out. One of the leading currents was *phenomenalism.* Phenomenalists rejected Kant's idea that phenomena are mere appearances of real objects. To the contrary, one may doubt the existence of things-in-themselves, but human perception of objects cannot be doubted. Subjective processes of perception have been designated as genuine foundations of all knowledge. According to Husserl (1931) everything could be doubted except the fact that human beings have *experiences*. Despite its professed opposition to Kant, phenomenology has perpetuated Kant's notion that the perceiving or experiencing subject was the center of the universe.

Ernst Mach offered a powerful support to the phenomenalist way of thinking. Mach did not accept Kant's idea of the thing-in-itself. According to Mach, the only thing human beings know are the sensory perceptions of the world. The so-called physical and mental phenomena, presumably outside our experience, are sheer metaphysical concepts. Science deals with observable facts, but the only observables are the reports of the sensory apparatus. "Pure experience," which is a cluster of sensations, is the sole foundation of science.

Mach is often believed to be a logical successor of Auguste Comte, but to a certain extent his system was a continuation of the radical idealistic metaphysics of George Berkeley. Berkeley wrote in 1710: "As several of these (ideas) are observed to accompany each other, they come to be marked by one name, and so to be reputed as one *thing*. Thus, for example, a certain color, taste, smell, figure, and consistence having been observed to go together, are accounted one distinct thing, signified by the name apple. Other collections of ideas constitute a stone, a tree, a book, and the like sensible things."

Berkeley doubted whether the perceived things exist. According to Berkeley:

> The table I write on I say exists; that is, I see and feel it; and if I were out of my study I should say it existed; meaning thereby that if I was in my study I might perceive it, or that some other spirit actually does perceive it. . . . For as to what is said of the absolute existence of unthinking things, without any relation to their being perceived, that is to me perfectly unintelligible. Their *esse* is *percipi;* nor is it possible they should have any existence out of the minds or thinking things which perceive them.

According to Berkeley *esse est percipi* (to be is to be perceived), for things that are not perceived by the perceiver are outside the scope of human knowledge. Mach "insisted on the direct discussion of observations and regarded scientific laws as the description of observations in the most economical way . . . Mach missed the point that to describe an observation that has not been made yet is not the same thing as to describe one that has been made; consequently he missed the whole problem of induction" (Jeffreys 1955:15–16).

Mach's philosophy resembles, to a certain extent, Wundt's ideas on introspection. According to Wundt, sensations, perceptions, etc. are directly, immediately given "inner experience" and cannot be doubted, thus psychology has to investigate "internal experience," such as sensations, thought, feeling, in contradistinction to objects of "external experience," that belongs to natural sciences (Wundt 1892; Brett 1912–1923).

According to Mach, "The world consists of colors, sounds, temperatures, pressures, spaces, times, and so forth, which is now we shall not call sensations nor phenomena, because in either term an arbitrary, one sided theory is embodied, but simply *elements*. The fixing of the flux of these elements, whether mediately or immediately, is the real object of physical research" (1903:208 ff.). Furthermore:

> Nature is composed of sensations as its elements. Primitive man, however, first picks out certain compounds of these elements—those namely that are relatively permanent and of greater importance to him. The first and older words are names of 'things.' . . . No inalterable thing exists. The thing is an abstraction, the name of a symbol. . . . Sensations are not signs of things, but, on the contrary, a thing is a thought-symbol for a compound sensation of relative fixation. Properly speaking, the world is not composed of 'things' as its elements, but of colors, tones, pressures, spaces, times, in short what we ordinarily call individual sensations. (1893:482 ff.)

Mach distinguished three steps in scientific inquiry. The sensory perceptions are the first and the only valid elements of knowledge. Logical operations and connections between the emirical observations are the second step. The formulation of simple and economical hypotheses that permit description and prediction is the third step.

Scientific theory should be directly derived out of observational data. Wrote Mach in 1872: "In the investigation of nature we have to deal only with knowledge of the connection of appearances with one another. What we represent to ourselves bhind the appearances exists *only* in our understanding" (p. 49). "If the hypotheses are so chosen that their subject never appeal to the senses and therefore also can never be tested . . . the investigator has done more than science, whose aim is facts, required of him—and this work of supererogation is evil. . . . In a complete theory, to all the details of the phenomena details of the hypotheses must correspond, and all rules for these hypothetical things must also be directly transferable to the phenomena" (1900:57).

Many German philosophers criticized Kant's ideas but the main opposition to Kant's philosophy is associated with the names of Auguste Comte and Herbert Spencer. Comte and Spencer demanded that philosophy renounce its status as the omniscient queen of knowledge. Comte's system of positive philosophy and Spencer's search for a synthesis of science introduced the philosophy of *positivism*, which represented a new approach to the acquisition of knowledge based on empirical research. The empiricism of the seventeenth and eighteenth centuries was more hortatory than factual; the new empiricists of the nineteenth century did not preach empiricism, but practiced it. The spectacular development of natural sciences in the nineteenth century has proven beyond doubt that scientists have little use for Kant's "Prolegomena." Instead of splitting philosophical hairs, researchers unraveled in a short time more knowledge than was done for centuries in philosophical armchairs.

Logical Positivism

The spectacular development of physical sciences in our century has forced philosophers to reexamine their conceptual tools. A group of philosophers and scientists including Moritz Schlick, Rudolf Carnap, Otto Neurath, Herbert Feigl, Philipp Frank, and Kurt Gödel was formed in Vienna in the early twenties. They called themselves *Der Wiener Kreis* (The Vienna Circle). Their contemporaries in Britain (and, in a way, associates) were Bertrand Russell and Ludwig Wittgenstein and later Susan Stebbing, R. B. Braithwaite, and A. J. Ayer; in Poland, Tadeusz Kotarbinski, Kazimierz Ajdukiewicz, J. Lukasiewicz, and later Taitelbaum-Tarski; in Germany, Hans Reichenbach, Richard von Misers, and later Carl Hempel. All of them could be classified as logical positivists.

According to logical positivism all scientific statements are either empirical-synthetical, which convey factual information, or formal-analytical definitions. Everything else was nonsense and most of the problems of yesterdays' philosophy were pseudoproblems (Kotarbinski 1929; Ayer 1959). Hume's words convey the spirit of the logical positivists.

> If we take in our hand any volume of divinity or school metaphysics, for instance; let us ask, Does it contain any abstract reasoning concerning quantity and number? No. Does it contain any experimental reasoning concerning matter of fact and existence? No. Commit it then to the flames: for it can contain nothing but sophistry and illusion. (Hume 1748)

Science was surging ahead notwithstanding old philosophers, and the new positivists have joined arms with the positive sciences. Following Wittgenstein's ideas expressed in his "Tractatus Logico-Philosophicus" (1921), the logical positivists emphasized the logical analysis of language. According to Carnap (1935), philosophy is to be replaced by the logic of science—that is, to say by the logical analysis of the concepts and sentences of the sciences, for the logic of science is nothing other than the logical syntax of the language of science.

Wittgenstein, Carnap, and their disciples believed that truthful elementary statements called *Protokollsätze* (protocol sentences) correspond to absolutely simple facts reminding one of Mach's elements, for statements were descriptions of observable events as experienced by the perceiving subject. The language of science advocated by the logical positivists was a system of "protocol sentences," that is sensory data reported in "observation sentences." Carnap wrote in 1932:

> Since the meaning of a word is determined by its criterion of application (in other words: by the relations of deducibility entered into by its elementary sentence-form, by its truth-conditions, by the method of verification), the stipulation of the criterion takes away one's freedom to decide what one wishes to "mean" by the word. If the word is to receive an exact meaning, nothing less than the criterion of application must be given; but one cannot, on the other hand, give more than the criterion of application, for the letter is sufficient determination of meaning. The meaning is implicitly contained in the criterion; all that remains to be done is to make the means explicit. (Carnap 1959:63)

Apparently the logical positivism was, in a way, a new version of Mach's philosophy. Freud was highly critical of this philosophy, as it will be explained later on.

Epistemological Issues

For centruies philosophers endeavored to exercise considerable influence if not control over scientific inquiry. A special branch of philosophy called *epistemology* was devoted to the analysis of the relationship between the scientist as an observer and the objects he observed. Epistemologists have scrutinized the conditions under which cognition takes place and maintain that they can determine the meaning of the term truth.

Immanuel Kant and John Locke suggested diametrically opposed answers. According to Kant, sensory evidence is no evidence at all, for whatever we perceive is colored by our a priori set mental framework called by Kant *Absoluter Geist*. According to Locke, there is no other source of knowledge but our sensory apparatus. However, Kant's categories of cognition led to Schopenhauer's epistemological solipsism, and Locke's empiricism was amply criticized by Hume, and Mach reduced the universe to a bundle of perceptions.

One may assume (and only assume) that what we perceive is or is not related to things as they are, and proceed from that point on or accept that the world is as perceived by us, or offer any other solution of combination of solutions.

An assumption that the universe is a cluster of perceptions leads to contradictions. Mach's world consists of Mach's perceptions; there is no way to prove (using Mach's perceptions) that the perceptions of other people are the same as his. One may follow Schopenhauer at this point and assume that the world is the way "I" see it. In such a case, any further pursuit of truth is futile.

Small wonder that Auguste Comte's system of positivism did not include any specific epistemological theory and William James' pragmatism related the idea of truth to realistic achievement. "My thinking," James wrote, "is first and last and always for the sake of my doing" (1890 2:333).

It seems that the only possible solution was that of *epistemological realism,* which assumes (and nothing more than assumes) that whatever exists, exists irrespective of whether someone perceived it or not. In other words, the earth was round before Copernicus, and America was where it is before Columbus' discovery.

Were the world a cluster of sensations, all sciences would have shrunk to psychology. However, such a panpsychologism, as developed by Berkeley or Mach, would render scientific psychology impossible. A psychologist would have to observe the way in which he observes, and so on, ad infinitum. He and his own perceptions would have become the sole source of information. There is no other way but to assume that other individuals exist independently of their perceivers. This is merely an assumption, for there is no empirical evidence for empirical evidence. Thus a *radical realism* is the only assumption free of inner contradictions.

A clinical psychologist observing the behavior of disturbed individuals assumes that he does not observe his own sensations, but the actions of other individuals. More psychologists may join in the process of observing, and all observers may observe the same phenomenon. Their observations can be checked against one another, and the knowledge thus acquired becomes objective, provided certain canons of scientific procedure are observed.

The first principle of epistemologic reason is that of *transcendent truth.* A proposition conveys transcendent truth whenever, and only when, its content corresponds to reality. The principle of transcendent truth requires that all propositions of empirical science be checked against reality.

Radical realism is not naive and does not assume an infallibility of human perceptions. Human perception can be improved by the use of scientific apparatus, by precision, by control, and by several other devices.

The aim of all these devices is to prove the correspondence between scientific propositions made by scientists and the objects of their inquiry. The objective of all these inquiries is to establish transcendent truth.

The logical principles, namely, identity, contradiction, and excluded middle (Cohen and Nagel 1934:181 ff), offer together the necessary overall principle. This principle of *immanent truth* means that scientific propositions must be free of inner contradictions and, within a given system, must not contradict one another. Thus, a scientific system is *formally* true whenever it is free of inner contradictions; then and only then does it meet the requirements of immanent truth. A scientific system is *empirically* true when it does not contradict the body of well-established empirical data and meets the requirements of *transcendent truth.*

Formal sciences, such as logic and mathematics, must meet the criterion of immanent truth only. Empirical sciences must meet both criteria (Wolman 1981, ch. 15).

Freud's studies represent the point of view of epistemological realism. Jung refuted these principles.

Empirical Generalizations

There are two levels of general statement in scientific inquiry. At the first level is the generalization from observable data: Archimedes' law of submerged bodies, the boiling point of water, the velocity of sound, or Ebbinghaus' law of forgetting. Empirical generalizations are open to empirical test by observation or experiment. They are validated by such an empirical test.

Empirical generalizations describe observable regularities of nature. Boyle's law of gases, Mendel's laws of genetics, the Weber-Fechner laws of psychophysics, Pavlov's laws of conditioning, Ebbinghaus' law of forgetting, and the Lewin-Zeigarnik law of incompleted tasks are examples of empirical generalizations, describing permanent relationships between observable events. They describe what happens and how it happens. They do not go beyond the "what" and "how" questions and do not answer the question "why" things happen. They may read as follows: "Whenever this and this happens, such and such is the sequence." Or: "Whenever an event p that belongs to class P takes place, an event q that belongs to class Q must take place." The relationship may or may not be one of temporal sequence. It may be stated in precise terms or in probabilistic terms, such as the dominant-recessive ratio in Mendelian laws.

All generalizations are statements dealing with classes or categories of bodies or events. The term "event" is used here (instead of the ambiguous term "phenomenon") to denote whatever happens to or with bodies.

Some logicians and philosophers of science call these empirical generalizations "empirical laws" (cf. Feigl 1951) or "experimental laws" (Nagel 1961). Both "law" and "experimental" are open to objection. It is true that many empirical generalizations have been derived from experimental studies, but one may develop empirical generalizations by using the observational method. Consider such a generalization as "All humans are mortal." Certainly, observation and induction alone offer valid evidence for such an empirical generalization. There are many empirical generalizations that require no experimentation; consider "All men are born," or "All men have two eyes." In this context, the term "generalization" is preferable to the term "law" because "law" may mean a certain relationship in nature that may be discovered by inductive or deductive method, whereas the term "generalization" is explicitly related to the scientific procedures of the research worker.

Experimentation is nothing more than observation under conditions planned and set by the experimenter. Planned experimentation offers a far superior and much more precise method of research than naturalistic observation. One may notice that the rails of a railroad expand on a hot day or that the highway surface melts under a scorching sun, but only experimentation can give a precise statement in regard to the impact of heat on physical bodies. One may notice that dogs salivate, but it took Pavlov's carefully planned experimentation to determine the laws of generalization, discrimination, and extinction.

Empirical generalizations, whether arrived at by observation or by experimentation, can be validated by controlled observation or experimentation. Empirical generalizations form the bedrock of any science. They may or may not express laws of nature; they may or may not be causal, e.g. Mill's division into types of laws. Empirical generalizations are milestones in the development of empirical science.

Psychology faces problems not encountered in physics or chemistry. Consider such a generalization, "Whenever a rock hit a windowpane, the glass broke." "Whenever p then q," or $p \supset q$; p stands for the first proposition, q for the second. Now, substitute an "iron door" for the "windowpane." Certainly, "iron" would not break. Thus, "Whenever p, then not $=$q," or $p \supset q'$; p stands for the proposition "A rock hit the iron door," and q' stands for "The iron door did not break."

Let us substitute a "man" for the "windowpane." The issue might be presented as follows: "The man's skin was scratched; there was some bleeding, swelling, and so forth." But this description is far from being complete, for a man, being a living organism, acts.

Were men simple organisms, their reaction would be either fight or flight. Men, however, react in various ways in addition to fight or flight. One man may start a verbal fight or a verbal flight. Another, in response to the blow, may rationalize, negotiate compensation, forgive, postpone revenge, blame himself, take any of several other courses of action, or possibly combine several. Thus, a general statement such as "Men hurt by rocks throw the rocks at the attacker," "Men hurt by rocks blame themselves," or "Men hurt by rocks forgive" is not true if "men" means "all men." Yet these may be true in regard to "some men"; furthermore, the same men may react differently, depending on who threw the rock, when, why, and what were their past experiences.

It has been frequently observed, that some children are thumb suckers. Investigation reveals several possibilities: Some children suck when the holes in the nipples are too big; they finish their milk too rapidly and go on to suck their thumbs. Some children suck when the holes in the nipple are too small; apparently, they get tired and give the bottle up, but suck their thumbs because they are hungry. When sucking the breast of an affectionate mother, sucking becomes a pleasurable experience; children suck their thumbs any time they are upset, as if seeking the soothing pleasure of sucking. Children who did not suck in childhood or were weaned too early suck their thumbs as if trying to recapture the missed gratification.

All four generalizations are apparently supported in some degree by empirical findings (cf. Halverson 1938; Isaacs 1933; Koch 1935; and others). Thus q (sucking the thumb) follows p_1 (deprivation in sucking). Obviously, p_1 contradicts p_2, and p_3 contradicts p_4. If p_1 is a, p_2 is not = a; if p_3 is b, p_4 is not = b. Yet q follows p_1, p_2, p_3, and p_4.

Even experimental animals do not act in an unequivocal way in support of one learning theory and against all other theories. A cursory review of the literature bears witness to this highly variated observable behavior. E. L. Thorndike's cat offered support to his master's thesis by prolonged trial-and-error learning (1911), but Guthrie's cat outdid him in loyalty to his master and learned in one-trial learning (Guthrie and Horton 1946).

In an experimental study of reaction to frustration conducted by Dollard et al. (1939), frustrated children turned aggressive. In another (Barker, Dembo, and Lewin 1941) frustrated children did not react with aggression but with regression. In experiments with face-to-face groups (Wolman 1960), the reaction to failure largely depended upon the objectives of the group.

Clinical observations yielded a rainbow variety of reactions. Some patients react to frustration with aggression or regression; others become very depressed. A schizophrenic patient, defeated in her efforts to set fire to the hospital, reacted with obvious relief. As she explained later, she realized she was not so dangerous as she had thought, if "one stupid doctor and one moronic attendant could overwhelm her." For a while she screamed "We are outnumbered," but later on she felt happy that finally someone "did not take her nonsense." Apparently, her reaction to frustration was ambivalent and included both protest and relief.

Pavlov's dog suffered a "nervous breakdown" when he was exposed to conflicting stimuli in an ellipse approaching a circle. Would a man suffer a nervous breakdown if he could not distinguish between his girl friend and her twin sister?

Psychological propositions are rarely simple. Physicists can remove air from a tube to find out whether bodies fall with the same speed in a vacuum. Human beings do not fit into evacuated tubes and their behavior is guided by a variety of reasons. They are inconsistent and change their minds; in similar situations they may react in dissimilar ways, and in similar ways in dissimilar situations. People marry for a variety of rational and irrational reasons; so do they love, hate, come together, and part. Simple and uniform generalizations cannot do justice to the complexity of human behavior. Complex hypothetical statements are necessary in psychology.

Evidence

Empirical sciences do not have a monopoly on truth, but scientific propositions (that is, statements) in contradistinction to all other areas of human communication are based on evidence. When a scientist says that water contains hydrogen and oxygen, or that depressive patients form one sixth of all admissions to mental hospitals in the United States, he must bring proof to his statements.

As described in the previous section, all empirical sciences present their findings in generalized statements which describe a class or a category of objects or events. Such empirical generalizations in psychology are derived from clinical and/or longitudinal observations of observable phenomena, such as thumb sucking, masturbation, impotence, mutism, aggression, sadism, and psychosomatic symptoms. These generalizations, analogously to experimental laws in physics, can be validated by controlled observations or controlled experimental or ex post facto experimental procedures. Even in physics a perfect validation may not be possible; probability calculus may be the only possible approximate proof.

The validity of these empirical generalizations depends on direct empirical evidence. Even in physical sciences the failure of an experimental study is not interpreted as the sole reason for the refutation of an experimental law. In some cases the experimental procedure is exposed to additional scrutiny; in other cases modifications of the empirical law are introduced.

There are three categories of empirical propositions in psychoanalysis. The first category deals with *observable* phenomena of manifest behavior, such as thumb sucking, masturbation, mutism, impotence, and suicide. The second category includes *introspectively observable* phenomena, such as depression, obsessive thoughts, dreams, pain, and pleasure. To the third category belong *unconscious processes*, such as compensation, repression, reaction formation, and all primary processes.

One should not be critical of the efforts to study the manifest aspects of behavior; yet it is necessary to state the observable data in a way that corresponds to the concept to be tested. In the case of psychoanalysis, several errors have been committed. No psychoanalyst could agree to the inaccurate description of repression given by Rosenzweig and Mason (1934) and other research workers. Also the definition of repression offered by Sears is at variance with Freud's ideas. According to Sears, repression itself is the blockage of the acts, either pure stimulus or contributory, which lead to the drive-satisfying goal.

Some experimentors tried to determine whether the word quinine, allegedly associated with unpleasant taste, would be less recalled than the word sugar, usually associated with sweet taste. Needless to say that such experiments have nothing to do with psychoanalysis. More sophisticated experiments conducted by Sears, Zeller, and other gave strong support to Freud's view on repression. Similar conclusions could be derived from the reviews of empirical research in psychoanalysis.

It is an open question whether experimental studies of overt behavior could serve as conclusive evidence for or against key psychoanalytic propositions. For ethical and/or practical reasons psychoanalytic experimentation must be limited to a small number of relatively simple symptoms. Psychoanalytic empirical generalizations describe conditions conducive to mutism, sadism, conversion, and psychosomatics, masturbation, impotence, bed-wetting, genocide, suicide, etc. None of these conditions can be enacted in an experimental setting.

Experimental research in psychoanalysis seems to be rather inconclusive: "The triviality of obtained differences in this field makes a most discouraging picture; and the coarseness of the experimental methods so far available for tapping the sensitive dynamics of repression does not augur well for the future," wrote Sears. Certainly controlled experimentation is the choice method, yet one may doubt whether the main psychoanalytic generalizations could be proven or disproven in an experimental design geared to the study of overt behavior.

Freud's reaction in 1934 to Rosenzweig's experimental study of repression was typical. Freud wrote: "I cannot put much value on these confirmations because the wealth of reliable observations on which these assertions rest make them independent of experimental verification. Still, it [experimental verification] can do no harm." Actually it does harm, for there is nothing more misleading than pseudoevidence.

Experimental studies can test certain manifest, observable data. Certain psychoanalytic propositions that deal with overt behavior could be experimentally produced. Those psychoanalytic propositions that deal with covert behavior, conscious and especially unconscious, present a separate problem.

Introspectively observable data, as a rule, evade experimental evidence.

Experimental study of dreams (Dement et al.) discovered highly interesting behavioral correlates of dreams, but did not disclose what the dreamer dreamed about (manifest dream) and certainly not what the dream represented (latent dream). Ultimately, even the manifest content of a dream is accessible to the dreamer only, and the nondreamer, either the analyst or the experimenter, can study the content of a dream and interpret it *after* it was communicated by the dreamer. The above-mentioned experimental studies offer an indirect support to Freud's observations.

Unconscious processes are not observable but inferable phenomena. They are not open to direct test, yet research can be planned in such a way that the unconscious phenomena would become evident as inevitable inferences from observable phenomena.

Explanation and Prediction

The need for a theory is not universal, self-evident, and everpresent. One can ascertain all telephone numbers in Manhattan without going beyond the empirical data clearly stated in the telephone directory. One can also count the number of windows in his house without using any theory at all.

The fact however remains that empirical data pertaining to objects and bodies and to what happens to them (facts) are not always as adequate as a telephone directory. Consider the production of cars or the building of houses. Why does gasoline make a car go? What does the steering wheel do to the wheels? What does the switch do to electric lights in a house? What does the window pane do to light and heat?

Descriptive statements, such as "all cars have a gas tank" or "bulbs go on at the turn of a switch" or "light goes through glass," are not of much help. Scientific research started when self-evident empirical data failed. The ancient Egyptians watched floods of the Nile; they knew all the facts and could *describe* them with great precision. But when the Egyptians tried to regulate the flow of water for irrigation, they had to seek connections between the facts. They had to reach *beyond* the observable data and seek *explanations*. In addition to the questions "what" and "how" things happen, they wanted to know *why* they happen. Prior to Freud, psychopathology was a collection of descriptive data, sort of a pile of bricks. Freud had added several new bricks (empirical data), but he has also developed an overall theory and built a house.

Apparently, there are three levels in scientific inquiry. The first is the ascertainment of facts and empirical proof of them; the second is abstraction, classification, and empirical generalization; the third is interpretation of empirical data or formation of a theory. A theory without facts is mere speculation, said David Hume. A pile of bricks is not a house and a collection of facts is not science, said Henri Poincaré (1902).

After having accumulated a great many detailed observations and drawing generalized statements, scientists go beyond the empirical data and endeavor to develop a system of hypotheses that interpret *why* things happen the way they do. This system of hypotheses is theory. A psychological theory, as any other theory, should enable scientists to explain and to predict.

As any other system of scientific propositions, a theory of human behavior must be free from inner contradiction. An explanation of facts must fulfill logical requirements and apply consistently the rules of implication and inference. A theory can be accepted, rejected, or modified on empirical or methodological grounds, but a self-contradictory theory cannot be truthful, and, therefore it is a priori unscientific. The principle of inner consistency and exclusion of inner contradiction, dervied from Aristotelian logic, is called *principle of immanent truth*.

The principle of immanent truth suffices in formal sciences, such as logic and mathematics, but does not suffice in empirical sciences. Empirical sciences deal with bodies and events, and explanatory theories must not only be self-consistent, but must also not contradict empirical evidence. The agreement with empirical evidence is called the *principle of transcendent truth*. (Dingle 1931; Wolman 1948, 1971; Nagel 1961).

Isaac Newton believed that physical theories must be derived from experimental data and maintained that he had deduced all his concepts and laws from experimental findings. Were Newton's statement of the relation between experimental observations and theory correct, Newton's theory would never have required any modifications nor would it contain theories which experiment does not confirm. Were theory merely deduced from facts, it would be as clear and as final as the facts are.

However, the experimental discoveries of Michelson and Morley in 1885 pointed to a fact which could not exist were Newton's theory entirely true. This discovery proved that the relation between experimental facts and theoretical assumptions is not what Newton believed it to be.

> When, some ten years later, experiments on radiation from black bodies enforced an additional reconstruction in Newton's way of thinking about his subject matter, this conclusion became inescapable. Expressed positively, this means that the theory of physics is neither a mere description of experimental facts nor something deducible from such a description; instead, as Einstein has emphasized, the physical scientist only arrives at his theory by speculative means. . . . Theories have to be proposed speculatively and pursued deductively with respect to their many consequences so that they can be put to indirect experimental tests. . . . For this reason, any theory is subject to further modification and reconstruction with the advent of new evidence that is incompatible . . . with its basic assumptions. (Heisenberg 1962:3–4)

A theory does not contradict empiricism. Had Newton, however, merely observed that apples fell (an empirical generalization), his role in the history of science would be rather negligible. Newton himself went beyond empirical findings; he asked *why* bodies fall, and the search for an answer led him to the formulations of a series of theoretical constructs, such as force, mass, and gravitation. Without these constructs, physics would have remained a pedestrian aggregation of observable facts. Analogously, without Freud, psychopathology would have remained a catalog of observable symptoms. Great scientists do not confine themselves to the discovery of new data; they create conceptual tools that help in fresh discoveries even when they themselves are unaware of their reaching beyond observable facts.

Some philosophers of science seem to believe that theory must be arrived at by an inductive method and empirical generalizations. This is not necessarily true, for neither Newton nor Einstein arrived at their bold theories through empirical generalizations. Certainly the history of psychology knows of intentionally deductive approaches to theory formation such as Hull's (1943) and Lewin's (1936).

Forming of Hypotheses

These two examples taken from the history of Freud's thought offer a good testing ground for theory formation in psychoanalysis. Freud himself embraced the philosophy of epistemological realism and was opposed to Kant's critical idealism and the Viennese logical positivism. Freud wrote in 1932 in the *New Introductory Lectures* about scientific thought as follows:

> It extends its interest to things which have no immediate obvious utility, it endeavors to eliminate personal factors and emotional influences, it carefully examines the trustworthiness of the sense perceptions on which it bases its conclusions. . . . Its aim is to arrive at correspondence with reality, that is to say with what exists outside us and independently of us, and, as experience has taught us, is decisive for the fulfillment or frustration of our desires. This correspondence with the real external world we call truth." (1932–33:233)

The assumption that the world exists independently from the observer is the essence of the epistemological realism. Whatever can be observed and/or experimentally tested, certainly should be observed and tested. This empiricist approach does not necessarily prejudge the status of unobservable events.

Freud subscribed neither to Kant's critical idealism nor to Comte's positivism. Freud's approach was that of empiricism, but not of its extreme and too radical version. The business of science is to study nature, whether it is or it is not observable.

Astronomy and chemistry offer a case in point. Astronomy proved the existence of celestial bodies before they were discovered. Similar things happened in the history of chemistry, where hitherto nonobservable chemical elements were discovered by nonempirical methods, and there is no reason for psychology to be *plus catholique que le pape*.

Freud's method of theory formation follows the principles evolved by such modern philosophers of science as Poincaré, Russell, and Einstein. The missing links between empirical data had to be postulated. Freud wrote:

> Every science is based upon observations and experiences arrived at through the medium of our psychical apparatus. But since our science had as its subject that apparatus itself, the analogy ends here. We make our observations through the medium of the same perceptual apparatus, precisely by the help of the breaks in the series of [conscious] mental events, since we fill in the omissions by plausible inferences and trans-

> late them into conscious material. In this way we construct, as it were, series of conscious events complementary to the unconscious mental processes. The relative certainty of our mental science rests upon the binding force of these inferences. (1938:36)

Scientific laws are established by inference.

> The processes with which psychology deals are in themselves just as unknowable as those dealt with by the other sciences, by chemistry or physics, for example; but it is possible to establish the laws which those processes obey and to follow over long and unbroken stretches their mutual relation and interdependences—in short, to gain what is known as an "understanding" of the sphere of natural phenomena in question. This cannot be effected without forming fresh hypotheses and creating fresh concepts. (ibid.)

And this is exactly how Freud developed his theory.

A scientific theory must also be *heuristically useful*. Each theory or system of hypotheses should be helpful in the future search for truth. A scientific theory may not be able to offer a final explanation of empirical data, but it has to be formulated in a manner that facilitates search for additional evidence. Whether a theory is composed of tentative hypotheses or of more positive explanations, room has to be left for future research and further elaborations. A useful scientific theory contributes to scientific predictability.

Theory Formation

A theory can be formed in several ways. It may evolve out of empirical data in a cautious induction; it may be deduced from arbitrarily set definitions and postulates. Validity of a theory does not depend upon the way it was arrived at but the way it serves empirical science, for a theory can be formed in any possible way and manner. Since theories use constructs and models, no theory, as Einstein stressed (1959), can be defined in operational terms.

In empirical science the concise test for a theory is its testability or correspondence to empirical data. "If a theory is to explain experimental laws," wrote Nagel, "it is not sufficient that its terms be only implicitly defined. Unless something further is added to indicate how its implicitly defined terms are related to ideas occurring in experimental laws, a theory cannot be significantly affirmed or denied and in any case is scientifically useless. . . . If the theory is to be used as an instrument of explanation and prediction, it must somehow be linked with observable materials" (1961:93).

This linkage between theory and observational data is often called *rules-of-correspondence* (p. 94ff). I shall try to prove that the *causal principle* is the main principle for psychological theories and the proper rule-of-correspondence between psychological theory and its empirical data. Theory means explanation, but there are several types of explanation.

Explanations can be deductive, probabilistic, functional or teleological, or genetic (Nagel 1961:20–26). Explanations correspond to some factual or formal relationships between empirical data or symbols representing them. Explanations seek unobservable links between observable phenomena. In certain research settings (to be discussed later) explanation stands for the intervening variable put between the observable dependent and independent variables.

One may distinguish four types of links or relationships: formal logical implications and inferences, mathematical functions and correlations, teleological connections, and causal relationships.

Logical implications (such as if p is greater than q and q is greater than r, then p is greater than r) are used in all sciences, but do not necessarily reflect real relationships. One can say logically that if Satan is stronger than the poor devil and the poor devil is stronger than the soul of a sinner, Satan is stronger than the unfortunate soul. All the former terms, although logically used, do not reflect much of reality.

Mathematical functions and correlations are exceedingly useful when backed up by empirical measurement or counting. It such cases they may reflect a true relationship. They may however be used on nonreal issues, such as irrational numbers or astrology or Herbart's mathematics. But even real correlations must not be confused with causation. For instance, "the longer the day, the shorter the night," is a perfect negative correlation, yet not a causal relationship. Or the higher the mercury in a thermometer, the higher the fever. Obviously mercury did not cause the illness and the fever.

Teleological connections have played a useful role in human behavior as much as they reflected the purposes, goals, and aims of human beings. Whenever the teleological principle has been elevated to the role of a general principle, it has somehow tended to ascribe human goals and purposes to nature. However, several psychological theories are based on teleology, among them those of William MacDougall and Alfred Adler.

The Causal Principle

The fourth type of relationship is the causal. At the present time philosophy of science raises serious doubts concerning the applicability of the causal principle in scientific inquiry.

Most of the difficulties of causation in modern physics stem from two sources, namely the quantum theory that cannot present its data in a cause-effect continuum and the relativity theory that has undermined the concept of temporal sequence. Thus, at least in microphysics, the concept of causation became useless and its transcendent truth questionable (Bergmann 1929; Braithwaite 1942; Cassirer 1956; Duhem 1954). "Observation," wrote Dingle (1931:88) "in Humean fashion teaches us only the facts of invariable succession. There cannot be any doubt that quantum theory has rendered the causal principle useless. The elaborate structure that the nineteenth-century science has built, seems to be dismantled and reduced to a pile of bricks."

Moreover, as Einstein pointed out: "At the present, we are quite without any deterministic theory directly describing the events. For the time being, we have to admit that we do not possess any general theoretical basis for physics, which can be regarded as its logical foundation." Contemporary physics cannot present its data in a deterministic continuum, but some physicists, among them Einstein himself, "cannot believe that we must abandon, actually and forever, the idea of direct representation of physical reality in space and time; or that we must accept the view that events in nature are analogous to a game of chance" (Einstein 1940).

Spinoza defined causation as follows: "Ex data cause determinata necessario sequitur effectus, et contra si null detur causa, impossibile est, ut effectus sequetur" (The effect follows necessarily from a given determined cause, and to the contrary it is impossible for an effect to follow if there is no cause (Axiom 3).

John Stuart Mill (1879: book IV, Ch. 3) maintained that the causal principle is "the ultimate major premise of all inductions." He defined cause as "the unconditional, invariable antecedent," and effect as the "invariable, certain, and unconditional sequence."

The element of necessary, inevitable temporal sequence is included in practically all definitions of causality. This necessary evolvement of effects out of causes inspired Spinoza to say that "Res aliqua nulla alia de causa contingens dicitur, nisi respectu defectus nostrae cognitionis" (We are calling a thing coincidental only because of deficiency of our cognition).

A similar idea was expressed by Laplace:

> We ought to regard the present state of the universe as the effect of its antecedent state and as the cause of the state that is to follow. An intelligence knowing all the forces acting in nature at a given instant, as well as the momentary positions of all things in the universe, would be able to comprehend in one single formula the motions of the largest bodies

> as well as of the lightest atoms in the world, provided that its intellect were sufficiently powerful to subject all data to analysis; to it nothing would be uncertain, the future as well as the past would be present to its eyes. The perfection that the human mind has been able to give to astronomy affords a feeble outline of such an intelligence. Discoveries in mechanics and geometry, coupled with those in universal gravitation, have brought the mind within reach of comprehending in the same analytical formula the past and the future state of the system of the world. All the mind's efforts in the search for truth tend to approximate to the intelligence we have just imagined, although it will forever remain infinitely remote from such an intelligence. (1820: Preface)

Hume raised doubts concerning the causal principle and deterministic nature of the world. Hume maintained that we *see* only temporal sequences (post hoc) and *assume* causal connections, *propterea hoc*. According to Hume, "We never can, by our utmost scrutiny, discover anything but one event following another, without being able to comprehend any force or power by which the cause operates, or any connection between it and its supposed effect. . . . All events seem entirely loose and separate. One event follows another: but we never can observe any tie between them" (1748:97). He also wrote: "We have no other notion of cause and effect but that of certain objects, which have been always *conjoined* together, and which in all past instances have been found inseparable" (1739–40:394).

As mentioned above, Kant solved the controversy of Locke versus Descartes by assuming that the world-as-it-is is not accessible to our sensory apparatus, and we view the world in terms of the perceiving human mind, and causality was one of those terms.

A hundred years ago Mach suggested substituting explanation by mere description, thus avoiding the causal principle altogether. Since, Mach wrote, "properly speaking, the world is not composed of 'things' as its elements, but . . . what we ordinarily call individual sensations" (1893:482), the business of science is not to interpose the nonempirical question "why" things happen but to describe "how" they happen.

Some contemporary physicists maintain that the causal principle is still the leading principle of natural science. According to Planck, "Physical science together with astronomy and chemistry and mineralogy are all based on the strict and universal validity of the principle of causality" (1932:147). And further on: "The last goal of every science is the full and complete application of the causal principle" (p. 158). Also Einstein took a definite stand on causation, assuming that its validity had not been finally abolished by the recent developments in theoretical physics.

It seems that the causal principle can be defined as *temporal sequence, ontologically necessary*, and *genetic*. Certainly, there are other, noncausal temporal sequences such as winter and spring. Furthermore, there are logically necessary, nontemporal sequences such as if a=b, and b=c, then a=c. There are certain ontologically necessary (yet not causal) sequences, such as day and night. There are genetic (productive) sequences such as a bud and a flower, a seed and a sapling, that are noncausal. Causality is a certain type of relationship that includes *sequence* in time, *ontological* necessity, and *genetic* elements (Wolman 1973).

The causal principle cannot be empirically proven. It may be postulated if it proves useful and helps the organization of factual data in a coherent system.

Causation in Psychology

Obviously, theoretical physics has substantial difficulties with causation, especially in quantum theory and the concept of time, but there is no reason for importing these difficulties into psychological research. In psychology the time sequence is a clearly definable issue, for there is a definite beginning for every human life and an irrevocable dead end to it. Whatever goes on between conception and death is well definable as prenatal life, infancy, childhood, adolescence, middle age, and old age. There cannot be any doubt that human behavior is a series of *temporal sequences*, but this cannot be said assuredly in regard to quantum mechanics.

Nor has *necessity* ever been questioned in empirical or clinical psychology. If you don't feed experimental rats, they become food deprived. When parents don't love their children, the children suffer emotional deprivation.

The problem of causality hinges also on another issue: "When we speak of something *causing* something else," wrote Morris R. Cohen, "we undoubtedly tend to attribute to the thing something analogous to human compulsion, something of muscular tension or the feeling of activity and passivity when we willfully push or are pulled contrary to our will. Such animism is out of place in modern scientific physics. . . . Technical and mathematical language, however, is surely if slowly replacing expressions of causal relations with mathematical functions or equations which are neutral to all anthromorphic hypotheses" (1931:224).

Behavioral sciences need not worry about this Humean distinction between succession and production. Hume fought against the unempirical idea of "forces" that allegedly produce change in nature. Nature cannot be bound by the fact that human action produces tools and weapons. It smacks of *anthropomorphism* to ascribe human talents to inanimate nature. But it is certainly a *reimorphism* to deprive human beings of human traits and prescribe laws of physics to humans (Wolman 1973).

Physics operates with the time-space construct; such a construct is inapplicable to psychological studies. There are no definable sequences in quanta. But in the realm of human life the time sequences are clearly established. Maxwell's fields "produce" but human life is full of unmistaken production, starting with genetics. Consider agriculture, industry, and creative writing. Consider population explosion, sex, fertilization, and heredity. Biological and behavioral sciences do not follow microphysics; they operate with different sets of data (Kaganov and Platonov 1961; Sarris 1968; Morganbesser 1967).

A psychological datum is *explained* when the datum is known and the *causes* are discovered. A psychological datum is *predicted* when the causes of an unknown datum are known and the *effects* are discovered. Explanation proceeds from the known effect to unknown causes; prediction proceeds from the known cause to the known effect. Explanation and prediction are two parts of the interpretative process; the first looks to the past, and the second into the future.

Psychological explanations and predictions are hardly unequivocal. Human predictions of the future and the explanation of the past are hampered by our inadequate knowledge of the present, imperfect knowledge of the past, and the inability to assess all factors that produce the future.

Theory Formation in Psychoanalysis

Psychoanalytic theories represent, at the first sight, a highly diversified series of conceptual systems hardly related to one another. However, all of them were related to clinical practice, and their roots lie in direct observation of patients' behavior, specifically their verbal behavior. Psychoanalysts of all schools have practiced and perfected the art of *listening*, and whatever they know is derived from the processes of communication and interpretation in contradistinction to most experimental studies based on visual observation and experimentation (Royce 1973).

Listening versus seeing is perhaps the main paradigm of psychoanalysis as compared to other branches of psychology. Experimental psychology developed out of two main sources: animal behavior and human neurophysiology. The first source is linked with such names as Charles Darwin, Lloyd Morgan, and Edward Lee Thorndike, the other is rooted in the studies of Johannes Miller, Ernst Heinrich Weber, Hermann Helmholtz, and others.

Introspection

Wundt followed the line of neurophysiological studies and called his main work *Principles of Physiological Psychology*. According to Wundt (1892), psychology differed from physiology in its subject matter, which is the "inner experience." This experience is two-sided. It is both the immediate experience of the experiencing person, as well as the experiencing person as a living organism who responds to external stimuli. Sensations are the elements of immediate experience; but sensations are aroused when a sensory organ is stimulated and sensory neurons conduct the excitations to the centers of the nervous system. Excitations of neurons and sensations are parallel phenomena.

Introspection, or self-observation, could hardly yield any scientific data as it was applied by earlier psychologists. Wundt was convinced that "the endeavor to observe oneself must inevitably introduce changes into the course of mental events—changes which could not have occurred without it, and whose unusual consequence is that the very process which was to have been observed disappears from consciousness." Therefore, he suggested combining introspection with experimentation, because psychological experiment "creates external conditions that look towards the production of a determinate mental process at a given moment. In the second place, it makes the observer so far master of the general situation, that the state of consciousness accompanying this process remains approximately unchanged. The great importance of the experimental method, therefore, lies . . . also and essentially in the further fact that it makes observation itself possible for us" (preface).

Wundt combined the observation of overt behavior with listening to subjects' communications reflecting their introspection, that is, self-observation.

The introspectionistic method was criticized on several counts. One can hardly be objective in regard to oneself, and introspectionistic psychology was accused of fostering subjectivism and producing dubious knowledge. Moreover, animals, children, and severely disturbed individuals could not be studied by the introspectionistic method.

The assault against introspection came from two sources; mental measurements and research in animal psychology. Mental measurements do not require introspection, yet the higher mental functions are being measured in a manner that enables one to predict future behavior and achievements. There is little doubt about the diagnostic and prognostic value of mental tests in educational, clinical, and industrial psychology although the theoretical discussion concerning the nature of intelligence still goes on. Intelligence tests served as ample evidence that highly important mental processes evade introspection and the best way of getting some insight into their nature is to put them to test. The "atmosphere" of mental tests was entirely different from the one prevailing in the laboratories of Wundt and Titchener. While Binet dealt with the problems of school systems and Yerkes faced a huge "human engineering" problem for the armed forces of the United States, Wundt's and Titchener's experiments were remote from life, and as arid as they were scientifically rigorous.

Animal psychology offered the second challenge to the introspectionist psychology. Charles Darwin's study on the *Expression of the Emotions in Man and Aminals* (1872) contained many truthful although not too precise observations. Darwin ascribed to animals emotional reactions similar to human feelings, and often used anthropomorphic expressions in talking about animals. Despite their shortcomings, Darwin's observations suggest in an irresistible way that there is quite a similarity between animal and human behavior and that human psychology can gain much by the study of animal behavior.

The Third Ear

The innovation introduced by Freud and wittingly or unwittingly accepted by all other analysts is related to a distinct disrespect for the value of human communication. Introspectionistic psychologists paid utmost attention to the content of the communication; Freud went beyond it.

The story is told of two psychoanalysts meeting on a street. When the first utters the conventional greeting, "How are you?" the second begins to wonder what indeed his colleague had in mind. When a patient says, "I hate you, doctor," the psychoanalyst tries to understand the hidden meaning of the communicated statements.

At this point, all psychoanalytic schools could be considered guilty of abandoning the empirical ground. Even radical behaviorists, such as J. B. Watson and B. F. Skinner, studied verbal behavior but did not go beyond observable auditory data. Skinner critized Freud for going beyond observable data and dealing with what Skinner called private events (Skinner 1956:79). But this is precisely what all psychoanalytic schools have been doing. The question is whether the "unobservables" are indeed beyond empirical science.

The Unconscious

Psychoanalysts disagree among themselves on a great many issues, but no one can be called a psychoanalyst unless he accepts the idea that a great part of psychological processes are partly or totally *unconscious*.

Prior to Freud, the term psychology was synonymous to the science of the consciousness. Psychologists studied consciousness, defined by introspectionists as the processes observed by the experiencing individual or "given in his inner experience." Processes that the individual was aware of and nothing else were considered the legitimate area of psychology.

Several phenomena, such as dreams, amnesias, and hypnotism, apparently belonging to the realm of psychology, were left out by psychology of the conscious. Yet it was rather arbitrary to dismiss the amazing mental phenomena of dreams, which captured the imagination of poets and prophets in ancient times. Far more difficult to overlook are the strange lapses of memory that occur in everyday life and in everyone's experience and the even more puzzling discontinuities in memory and the conscious caused by emotional disorders.

Interoceptive Conditioning

The unconscious is not a theory; it is an empirical fact. In addition to the evidence adduced by Freud, to be described later on, Soviet research in interoceptive conditioning offered unexpectedly strong support to Freud's concept of unconscious. For a while psychoanalytic concepts were taboo in Russia, but the latest developments bear witness, if not to a change of mind, at least to a change in factual information and, perhaps, to a beginning of a rapprochement. Teplov pointed to the fact that human psychic processes are made conscious by being reflected in the second signal system. Not all mental processes are reflected there, thus not all are conscious (1933:260ff.).

The neurophysiological tradition has always been strong in Russian research; even Sechenov and Pavlov did not insist on the exclusive study of overt behavior. Certainly they did not share the introspectionist aversion to unconscious processes or the behavioristic dislike for conscious processes.

Several experimental studies conducted by Bykov and others reported conditioning i.e., a mental process not accompanied by "subjectively perceived sensation," i.e., consciousness. Ayrapetyantz et al. (1952) and several other Soviet workers stressed the fact that, on a low ontogenetic level, conditioning processes are unconscious; so are many interoceptive conditioning processes in humans. For a while unconscious processes were inaccessible to experimental research. Present-day experimental research in interoceptive conditioning in men is de facto experimentation in unconscious processes.

"Experimental Subjects"

When one reads the works of Freud, Jung, Adler, Fromm, and Sullivan one wonders whether there is any common denominator besides the concept of unconscious. As mentioned above, Freud followed the empiricist tradition of Darwin, Adler was influenced by the neo-Kantian philosopher Vaihinger, and Jung's inspiration came from oriental religious philosophies.

However, despite the philosophical differences, the patients were the main source of knowledge common to all psychoanalytic thinkers. These auditory suppliers of information come from a particular type of population.

No one knows the ratio of disturbed people in a given population, nor is it possible to assume that the patients seen by Freud or Sullivan are representative of all types of mental disorder. Thus no psychoanalytic theory can claim absolute universality and no psychoanalytic theory can explain *all* aspects of human behavior.

Moreover, as psychoanalysis developed out of direct interaction between patients and psychoanalysts, the effect of participant observation cannot be overlooked. Compare, for instance, Freud's and Sullivan's ideas on the nature of schizophrenia. Freud saw patients in his office, Sullivan in a mental hospital. The reclining position of the patient and the sparse communication from the silent psychoanalyst facilitated transference phenomena. Freud made transference the cornerstone of his therapeutic method. However, when psychotic patients were asked to recline with the psychoanalyst sitting behind the couch and watching, silence was perceived as a rejection and the invisible analyst became a threatening figure. The patients withdrew even more and did not dare to communicate the true feelings. Such a behavior gave the impression of narcissistic withdrawal and lack of transference feelings.

Sullivan worked with patients moving back and forth and acting out their feelings. He saw patients in interpersonal relations, and he did not fail to notice the socially induced changes in their behavior. The interaction became the clue to the understanding of the psychotic personality

(Sullivan 1947, 1953). Most probably what has been observed in a participant observation was not the patient as an isolated entity but the patient in interaction with the therapist, a *psychosocial* field situation.

Theory Construction

Some psychologists believe that there is but one way of theory formation, starting with observation of empirical phenomena and proceeding toward most general statements. Simple observation can present human actions on a stimulus response continuum, the stimulus representing the totality of actions done by the environment and the response standing for whatever happened with the organism as a result of the stimulation. However, even a superficial observation shows that the response of the organism is more than a response to the stimulation: a hungry rat and a satiated one will respond differently to the sight of food. Credit has to be given to R. S. Woodworth for modifying the S-R formula into S-O-R, in which O stands for the organism, and it is understood that R is a result of what was produced by S and O together. Undoubtedly it was a substantial improvement in the presentation of psychological events.

Tolman refined Woodworth's proposition and presented the S-O-R formula in a continuum of experimental variables in which S is the independent, R the dependent, O the intervening variable.

This distinction between intervening variables and hypothetical constructs introduced by MacCorquodale and Meehl (1948) has been further elaborated by Feigl (1948), who preferred to call the constructs existential hypotheses. These "fill out" the space assigned to intervening variables, as, e.g., the hypothetical constructs of the theory of genes "fill out" Mendel's conceptions of heredity, or the intervening variable of temperature is being identified with the existential hypotheses of microstates of molecules. The existential hypotheses are introduced "on the basis of some new and heterogeneous area of evidence." At the price of existential hypotheses we achieve a reduction (often a very considerable one) of the hitherto unreduced dispositional concepts or intervening variables.

One may introduce any logical constructs at any level of explanation, provided they are clearly defined and serve a useful purpose. Watson's subvocal speech, Allport's trait, Freud's unconscious are intended to report facts which cannot be directly observed. They cannot be presented as intervening variables; they are inferables.

"Unconscious has more and more been made to mean a mental province rather than a quality which mental things have," said Freud. The unconscious as a "mental province" is a logical construct of mind. Most logical constructs are "methodological bridges" used for interpretation of behavior. Freud's model of personality is composed of several constructs: id, ego, and superego.

Some students of psychoanalytic theories seem to believe that these theories developed out of empirical generalization. Were this true, Freud's or Horney's system should be rewritten in a more or less following position: "Most of the patients I treated during a given period were obsessed with . . ." or "All my patients displayed the following symptoms."

Nothing could be more alien to the psychoanalytic way of thinking than the above. Some psychoanalytic propositions are generalizations, for instance, that the superego usually develops at the phallic stage. However, Freud's concept of superego or Sullivan's self-system are not empirical generalizations. They are *bona fide* logical or theoretical constructs.

Some experimental psychologists trained in using laboratory equipment believe in the need for presenting their data in operational definitions. As long as operationism is applied to experiments with observable data, it is just a new slant of empiricism. It cannot be applied to introspective data, nor to unobservable, unconscious phenomena. An effort to define operationally the unobservable inferables such as regression, reaction formation, or transference is doomed to failure because the only observables are acts of overt behavior. Least justified are operational definitions of theoretical concepts. What kind of operations could one perform that would lead to an operational definition of ego, libido, or superego? These terms are theoretical constructs, i.e., nonempirical parts of Freud's system. Theoretical constructs are constructs and not facts, and as such they are neither true nor false. Constructs are scientific tools, useful or useless but not factual data. They can and should be modified whenever empirical data or methodological convenience suggest such a modification.

Psychoanalytic propositions are open to criticism. They are not factual statements or religious dogmas. Neither are the propositions developed in any other experimental science. Consider learning theories. The issues of S–R theories are so vague at critical points that "they cannot be shown to be false. . . . We may begin to understand why some psychologists suspect that neither of these hypotheses is testable," concluded Ritchie in an analysis of learning theory (1965). Ritchie pointed to the fact that despite decades of experimental research neither Hull's school nor Tolman's school nor any other learning theory was capable of disproving the competitive schools.

It is true that psychoanalytic propositions may bring logicians to despair, because of the irrationalities they describe or explain. But it is not that psychoanalysis is irrational; human behavior is irrational. It is easy to develop arbitrary systems that agree with one's logic. Psychoanalysis as an empirical discipline is faithful not to any peculiar logical system but to the irrational subject matter—human behavior. It is a fact that fixation can be caused by either over- or undergratification; it is also true that a father's death may cause several types of reactions. Psychoanalysis has to study human reactions, irrational as they are, and not as obvious as one may wish them to be. Freud believed that the least rigid hypothesis will be the best in regard to obscure and inaccessible regions of the human mind.

There are three criteria for the rejection of a theory. The first is the criterion of inner consistency or immanent truth. The second criterion is the agreement with the established body of empirical findings, or the transcendent truth. Third is the extent to which a theory is more explanatory and more knowledge producing than a competing theory.

As far as transcendent truth is concerned, psychoanalytic theory is, like any other theory, a set of constructs and hypotheses explaining empirical facts. It cannot be proven or disproven by direct empirical evidence. Thoery has to be open for deductions that are open to empirical test. The entire field of psychopathology is open to such a test. Empirical test cannot refute theory directly; however, it may imply necessary modifications.

It is true that Freud often used pictorial language, for ego or superego or anxiety or libido or Eros are terms that do not describe any observable phenomena. Also physics uses nonexperimental terms such as neutrino or electron or field. Theoretical terms are formulated in a language different from the empirical terms and are communicated in terms alien to experimental procedure. Einstein was opposed to Bridgman's operationism in theoretical physics but some psychologists insist on "operational definitions" of psychoanalytic theory. Such a definition cannot be true or false; it is simply nonsense.

A theory or a theoretical construct or model is scientifically useful when experimental or observational laws can be deduced from them. These experimental (in physics) and observational laws (in psychology) should be in better agreement with observable data than any other laws, or there is no use in introducing them. In other words, a hypothesis has to be rejected when it is more at variance with observables than other hypotheses. The criterion of transcendent truth cannot be applied too rigorously in physics; it is indeed hard to believe that this principle can be applied in psychoanalysis more rigorously than in physics.

As far as inner consistency or the immanent truth is concerned, an optimum of inner consistency is desirable, but perfection is unattainable even in mathematics. Psychoanalytic theory is free from logical contradictions even when it describes self-contradictory human behavior. It is logically true that the term love is the opposite of the term no-love. It is empirically true that the very same individual may experience love and no-love toward the same person, may fear and not fear, may admire and despise, may be happy in pain and miserable in victory at one and the same time. Psychoanalysis sticks to empiricial facts and not to arbitrary rules.

BIBLIOGRAPHY

Ayer, A. J., ed. 1959. *Logical Positivism*. Glencoe, Ill.: Free Press.

Ayrapetyantz, E. 1952. *Higher Nervous Function and the Receptors of Internal Organs*. Moscow: USSR Academy of Sciences.

Barker, R. G., T. Dembo, and K. Lewin. 1941. Frustration and regression: An experiment with young children. *University of Iowa Studies in Child Welfare,* 18:1–134.

Bergmann, H. 1929. *Der Kampf um das Kausalitätsprinzip in der jüngsten Physik*. Leipzig: Barth.

Berkeley, G. 1710. *Principles of Human Knowledge*. Oxford: Oxford University Press, 1901.

Braithwaite, R. B. 1942. *Scientific Explanation*. Cambridge: Cambridge University Press.

Brett, C. S. 1912–23. *History of Psychology,* 3 vols. London: Allen and Unwin.

Carnap, R. 1935. *Philosophy and Logical Syntax*. London: Kegan Paul.

—— 1959. *The Elimination of Metaphysics through Logical Analysis of Language*. In A. J. Ayer, ed., *Logical Positivism*. Glencoe, Ill.: Free Press.

Cassirer, E. 1956. *Determinism and Indeterminism in Modern Physics*. New Haven: Yale University Press.

Cohen, M. R. 1931. *Reason and Nature*. New York: Harcourt and Brace.

Cohen, M. R. and E. Nagel. 1934. *An Introduction to Logic and Scientific Method*. New York: Harcourt, Brace.

Comte, A. 1830–42. *Cours de philosophie positive*. Paris: Baillère, 1864.

Descartes, R. *Philosophical Works*. 2 vols. Cambridge: Cambridge University Press, 1931.

Dingle, W. 1931. *Science and Human Experience*. London: William Norgate.

Dollard, J., N. E. Miller, L. W. Doob, O. H. Mowrer, and R. R. Sears. 1939. *Frustration and Aggression*. New Haven, Conn.: Yale University Press.

Duhem, P. 1954. *The Aim and Structure of Physical Theory*. Princeton, N. J.: Princeton University Press.

Einstein, A. 1940. The fundaments of theoretical physics. *Science,* 91:487–492.
—— 1959. Reply to criticism. In P. A. Schilpp, ed., *Albert Einstein: Philosopher Scientist,* pp. 663–688. New York: Harper.
Feigl, H. 1957. Principles and problems of theory construction in psychology. In W. Dennis, ed. *Current Trends in Psychological Theory.* Pittsburgh, Pa.: University of Pittsburgh Press.
Feyerabend, P. K. 1963. How to be a good empiricist. In B. Baumrin, ed., *Philosophy of Science.* 2 vols. New York: Wiley.
Fox Bourne, H. R. 1876. *The Life of John Locke.* 2 vols. New York: Harper.
Freud, S. 1932. *New Introductory Lectures on Psychoanalysis.* New York: Norton, 1933.
—— 1938. *An Outline of Psychoanalysis.* New York: Norton, 1949.
Guthrie, E. R. and G. P. Horton. 1946. *Cats in a Puzzle Box.* New York: Rinehart.
Halverson, H. M. 1938. Infant sucking and tensional behavior. *Journal of Genetic Psychology.* 53:365–430.
Hegel, G. W. F. 1807. *Phenomenologie des Geistes.* Berlin: Duncker and Humblot, 1841.
Heisenberg, W. 1962. *Physics and Philosophy.* New York: Harper and Row.
Hull, C. L. 1943. *Principles of Behavior.* New York: Appleton-Century-Crofts.
Hume, D. 1739–40. *A Treatise on Human Nature.* London: Longmans, 1874.
—— 1748. *An Enquiry Concerning the Human Understanding.* Oxford: Clarendon, 1894.
Husserl, E. 1931. *Ideas: General Introduction to Pure Phenomenology.* London: Allen and Unwin.
Isaacs, S. 1933. *Social Development in Young Children.* London: Kegan Paul.
Jeffreys, H. 1955. *Scientific inference,* 2nd ed. Cambridge: Cambridge University Press.
Kaganov, V. M. and G. V. Platonov. 1961. *The Problem of Causality in Contemporary Biology.* Moscow: USSR Academy of Sciences.
Kant, I. 1781. *Critique of Pure Reason.* London: Macmillan, 1929.
Koch, S. 1935. An analysis of certain forms of so-called nervous habits in small children. *Journal of Genetic Psychology,* 46:139–170.
Kotarbinski, T. 1929. *Elements of the Theory of Knowledge, Formal Logic, and Methodology of Science* (Polish). Lwow: Atlas.

Kuhn, T. 1962. *The Structure of Scientific Revolutions*. Chicago: University of Chicago Press.

Laplace, P. S. 1820. *Théorie analytique des probabilités*. Paris: Alcan.

Lewin, K. 1936. *Principles of Topological Psychology*. New York: McGraw-Hill.

Locke, J. 1690. *An Essay Concerning Human Understanding*. Oxford: Oxford University Press. 1894.

Mach, E. 1900. *History and Roots of the Principle of Conservation of Energy*. La Salle, Ill.: Open Court, 1911.

—— 1903. *Popular Scientific Lectures*. Chicago: Open Court.

—— 1893. *The Science of Mechanics*. Chicago: Open Court, 1960.

MacCorquodale, K. and P. E. Meehl. 1948. On a distinction between hypothetical constructs and intervening variables. *Psychological Review*, 55:95–107.

Mill, J. S. 1843. *A System of Logic*, 8th ed. London: Longmans, Green, 1879.

Morgenbesser, S., ed. 1967. *Philosophy of Science Today*. New York: Basic Books.

Nagel, E. 1961. *The Structure of Science*. New York: Harcourt, Brace.

Pavlov, I. P. 1955. *Selected Works*. Moscow: Foreign Languages Publishing House.

Planck, M. 1932. *Where Science Is Going?* New York: Norton.

Poincaré, H. 1902. *La science et l'hypothèse*. Paris: Flammarion.

Royce, J. R. 1973. The present situation in theoretical psychology. In B. B. Wolman, ed., *Handbook of General Psychology*, pp. 8–21. Englewood Cliffs, N.J.: Prentice Hall.

Sarris, V. 1968. Zum Problem der Kausalitat in der Psychologie. *Psychologische Beiträge*, 10(2):173–186.

Sears, R. R. 1943. *Survey of Objective Studies of Psychoanalytic Concepts*. Social Science Research Council.

Schopenhauer, A. 1818. *The World as Will and Idea*. New York: Scribner, 1923.

Skinner, B. F. 1953. *Science and Human Behavior*. New York: Macmillan.

Spinoza, B. 1677. *Ethics*. London: Bell, 1919.

Stern, W. 1924. *Person und Sache*. Leipzig: Barth.

Sullivan, H. S. 1953. *Interpersonal Theory of Psychiatry*. New York: Norton.

Thorndike, E. L. 1911. *Animal Intelligence: Experimental Studies*. New York: Macmillan.

Wittgenstein, R. 1922. *Tractatus logico-philosophicus*. London: Kegan Paul.

Wolman, B. B. 1948. *Prolegomena to Sociology* (Hebrew). Jerusalem: Kiryat Sefer.

—— 1960. Impact of failure on group cohesiveness. *Journal of Social Psychology*, 51:409–418.

—— 1968. Immanuel Kant and his impact on psychology. In B. B. Wolman, ed., *Historical Roots of Contemporary Psychology*. New York: Harper and Row.

—— 1971. Does psychology need its own philosophy of science? *American Psychologist*, 26:877–886.

—— 1973. Concerning psychology and the philosophy of science. In Wolman, ed., *Handbook of General Psychology*, pp. 22–48. Englewood Cliffs, N. J.: Prentice-Hall.

—— 1981. *Contemporary Theories and Systems in Psychology*. New York: Plenum.

Wundt, W. 1874. *Principles of Physiological Psychology*. London: Macmillan, 1940.

[TWO]
Cultural-Historical Perspectives

WERE Freud's times ready to accept his theories? Why has Freud met with so much opposition and criticism? What inspired the American psychologist Dunlop Knight to write the book *Psychoanalysis, Mysticism or Science?* Why even today do most academic psychologists reject psychoanalytic theories?

The waves of history of civilization move with a rather uneven speed. Centuries went by with little visible progress in science and technology, but the nineteenth century evolved with a hitherto unheard of speed. Physics and chemistry, biology and medicine, physiology and neurology made tremendous strides. Even psychology, the sleeping beauty enchanted by metaphysics, woke up in 1879 and established the first experimental research center. A formidable wall of self-respectability surrounded the proud centers of scientific research.

Progress in Physics

In 1819 Hans Christian Oersted (1791–1867) discovered electrolysis and electromagnetic induction and pioneered the electromagnetic theory of light. James Clark Maxwell (1831–1879) published the *Treatise on Electricity and Magnetism* (1873) which for decades has been called "The Electricians' Bible." In 1880 Thomas Edison invented the electric bulb, thus harnessing both nature and science for the service of man.

Maxwell's electromagnetic theory was subsequently experimentally tested and further developed by Heinrich Hertz (1857–1894) and Hendrik Anton Lorentz (1853–1928). At the crossroads of the nineteenth and twentieth centuries the electromagnetic theory of light was well established, and the famous Michelson-Morley experiment in 1887 heralded the beginning of new and bold physical discoveries, among them the theory of electrons, gaseous ions, radioactive rays, and, eventually, the quantum and relativity theories.

The development of thermodynamics had a most significant impact on the growing physiological studies and the budding psychological science. The study of liquefaction of gases, the discovery of osmosis (Dutrochet 1827), and the establishment of the principle of conservation of energy opened new vistas and broke the ground for the idea that one type of energy can be transferred into another. This idea enabled Freud to think of libido as a sort of derivative from physical energy.

The second half of the nineteenth century made people believe that the strictly scientific experimental method would open all doors to mankind's uninterrupted progress. In 1871 Mendeleev discovered the periodic table of chemical elements. In 1895 Marconi built the telegraph.

The progress of physical sciences has stimulated psychology and exercised an irresistible influence on psychological research and speculations. Even the incredible Mesmer, in his efforts to develop a method of influencing people, followed physical models. Mesmer's *bouquet* was an imitation of Leyden's jar. In the eighteenth and nineteenth centuries, experiments with magnetism and electricity were very much in vogue, and the common belief was that electricity is some sort of fluid that can be stored and conveyed through physical contact between people. In 1745 Ewald von Kleist filled a bottle with water and put in the water a nail attached to a gun barrel. When he applied the nail to an object and touched the object, he felt a powerful shock. Kleist concluded that "the human body must play some part in the effects."

A few months later Pieter van Musschenbroek in Leyden repeated the experiment. This Leyden experiment "gave a great boost to electricity which henceforth became a general subject of conversation. All the electricians of Europe tried to repeat it and to study its significance," wrote Abbé Mollet (quoted after Taton, 3:476).

Biology

The nineteenth century was also the century of biology. Charles Darwin (1809–1892), Karl von Nageli (1817–1891), Jean Baptiste de Lamarck (1744–1829), Herbert Spencer (1820–1903), Alfred Russel Wallace (1823–1913), Thomas H. Huxley (1825–1895), August Weismann (1834–1914), Gregor Mendel (1822–1899), Louis Pasteur (1822–1895), Ernest Haeckel (1834–1919), Hugo de Vries (1848–1935), and many others brought about far-reaching changes in human concepts of nature.

Lamarck in his work *Philosophie Zoologique* (1809) introduced the theory of evolution. He maintained that new needs lead to the development of new organs, and the acquired changes are inherited by the subsequent generations. The biblical view of immutable biological species being created by the Lord on one day and put at the service of man was to be more and more challenged by biological studies. The idea of mankind being no more than just another biological species driven by innate forces has gradually pervaded all centers of learning around the world.

In 1798 Robert Malthus wrote:

> I think I may fairly make two postulates. First, that food is necessary to the existence of man. Secondly, that the passion between the sexes is necessary and will remain nearly in its present state.

And furthermore:

> I say that the power of the population is indefinitely greater than the power of the earth to provide subsistence for man . . . [for] population unchecked increases in geometrical ratio, subsistence only increases in an arithmetic ratio.

Charles Darwin described Malthus' influence as follows:

> In October 1838 . . . I happened to read for amusement Malthus on Population, and being well prepared to appreciate the struggle for existence which everywhere goes on from long continued observation of the habits of animals and plants, it at once struck me that under these circumstances favorable variations would tend to be preserved, the unfavorable ones to be destroyed. The result of this would be the formation of a new species. Here then I had at last got a theory by which to work. (Darwin 1892:7)

Twenty years later, Alfred Russel Wallace (1823–1913) arrived at a similar conclusion. He wrote in his autobiography:

> In February 1858 . . . something led me to think of the positive checks described by Malthus in his Essay on Population, a work which I had read several years before, and which had made a deep and permanent impression on my mind. These checks—war, disease, famine and the like—must, it occurred to me, act on animals as well as man. . . . There suddenly flashed upon me the idea of the survival of the fittest—that the individuals removed by these checks must be on the whole inferior to those that survived.

After Lamarck, Darwin, Wallace, Weismann, and others, no scientist could believe in the Bible story of creation. Man was no longer the king of the universe and his organism and its functions were to be exposed to an objective, matter-of-fact inquiry.

Neurophysiology

Viewing humanity as a biological species brought about spectacular achievements in the world of neuroscience. François Magendie (1783–1855) and simultaneously Charles Bell (1774–1842) discovered that the posterior spinal nerve roots and the front motor nerve fibers were sensory, composed of dorsal roots and spinal ganglia, while the ventral roots of the spinal cord include motor nerves only. Magendie and Bell found that conduction goes one way, the stimulus being conducted along one path from the nerve endings in a sensory organ to the center of the nervous system and response goes from the center along another path. The discovery of this *law of forward conduction* led to the stimulus-response, or *reflex-arc* theory of behavior.

Johannes Müller (1801–1858) arrived at far-reaching conclusions concerning the sensory process. If perception depends on the nerve endings of the sensory apparatus, it cannot be totally determined by the external stimulus, and the nature of the sensory apparatus must affect the content of perception. Müller's law of *specific energies of sensory nerves* reads as follows:

> External agencies give rise to no kind of sensation which cannot also be produced by internal causes exciting changes in the condition of our nerves . . . (e.g. visual sensations) are perceived independently of all external exciting causes. . . . The excited condition of the nerve is manifested, even while the eyes are closed, by the appearance of light or luminous flashes which are merely sensations of the nerve. (Müller 1838–1842)

At the same time Pierre Flourens (1794–1867) removed the cerebral lobes of the brain of pigeons. He studied the loss of a certain part of the brain on behavior, thus laying the foundation for cerebral *localization* theories. Flourens maintained that the brain lobes perform all the functions of perceiving, memorizing, reasoning, and decision making, while the cerebellum coordinates motor behavior.

Claude Bernard (1813–1878) developed experimental techniques in the study of the human organism. He maintained that experimental method which was so productive in physics and chemistry should be applied to physiological processes. Bernard revolutionized medical science and stimulated experimental research in physiology, endocrinology, and neurology.

Helmholtz

The second part of the nineteenth century witnessed a great many discoveries in neurophysiology. Müller and his disciples Brücke (Freud's teacher), du Bois-Reymond, Helmholtz, and Ludwig formed sort of a scientific alliance advocating a radical brand of monistic materialism. The disciples of the above-mentioned Hermann von Helmholtz (1821–1894) formed in 1847 a group, the Helmholtz School of Medicine. The 1847 group exercised considerable influence in their attempts to reduce physiology to chemistry and physics, to promote strict causality, and to encourage radical empiricism and experimentation. The 1847 group was vastly successful in experimental physiology, temporarily successful in putting across mechanism against vitalism, and reasonably successful in popularizing applied physics and chemistry as techniques in physiological research. But Ludwig, Helmholtz, du Bois-Reymond, and Brücke had for a short time held most of the German physiology to a much bolder task, that of dissolving physiology into physics and chemistry. It was this task which Ludwig himself admitted proved "much more difficult than we had anticipated" (Cranefield 1957:423).

As mentioned before, Sigmund Freud studied under Brücke, and Brücke's ideas influenced Freud's initial concepts described in the unpublished by Freud "Project for Scientific Psychology" (posthumously published in 1953, Freud's *Collected Works,* 1:283–398).

Fechner

Freud was also greatly influenced by Gustav Theodor Fechner (1801–87). His *Elements of Psychophysics* (1806), broke the ground for a scientific psychology. According to Fechner, the understanding of the process of perception bridged over the gulf between the physical and the mental worlds. The *stimuli* coming from without the organism represented the physical world; the responses coming from within, the sensations, were mental. Hence, Fechner's approach was called *psychophysics.*

Fechner used experimental methods in research; his data were quantitative; the mathematical relations between physical stimuli and mental sensations were clearly stated. As stimuli increase in geometrical progression, sensations increase in arithmetical progression. This was the generalization of Weber's findings which became known as Weber's law.

Fechner also elaborated upon the concept of threshold. The *initial threshold* was the minimal intensity of a stimulus necessary for the stimulus to be perceived. The *differential threshold* was the minimum of increase or decrease in intensity of the stimulus necessary for the perception of these decreases or increases.

Besides his careful studies concerning thresholds, which are an important contribution to physiology and psychology, Fechner developed three psychophysical methods: (1) the method of *limits,* in which the stimuli in an experiment are ordered either in increasing or in decreasing intensity; (2) the method of *production,* in which the subject controls the stimuli according to the experimenter's instructions; (3) the *constant* method, in which stimuli act upon the subject in an irregular order.

Fechner's pain and pleasure theory had a direct impact on Freud. Freud quoted Fechner in *Beyond the Pleasure Principle,* as follows: "In so far as conscious impulses always have some relation to pleasure or unpleasure, pleasure and unpleasure too can be regarded as having a psycho-physical relation to conditions of stability and instability." In 1920, Freud wrote: "We have decided to relate pleasure and unpleasure to the quantity of excitation that is present in the mind but is not in any way "bound"; and to relate them in such a manner that unpleasure corresponds to an *increase* in the quantity of excitation and pleasure to a *diminution.*"

Helmholtz's theory of conservation of energy and his neurophysiological research in vision and hearing opened new vistas for the budding psychological science. Freud's theory of libido and cathexis were based on the principle of conservation of energy applied to neurophysiological and, later on, to psychological processes.

Helmholtz's experimental research of the speed of neural conduction was a milestone in the history of physiological psychology. All studies of "reaction time" have been a continuation of Helmholtz's research, and Freud's initial ideas of thwarting and blocking of discharge were, in a way, patterned after Helmholtz's scientific model (Bernfeld 1944).

Psychiatry

In the nineteenth century there was no clear-cut distinction between the two branches of medical science, neurology and psychiatry. In the second part of the nineteenth century, psychopathology and psychiatry were dominated by a clearly organic, neurological point of view. In 1845 Wilhelm Griesinger published a textbook of psychiatry that was used for decades as a standard-bearer of the science of psychiatry all around Europe. Griesinger believed that mental disorders were brain diseases. The only psychological theory that Griesinger accepted was Herbart's mechanistic-materialistic system, applied toward the understanding of neurophysiological processes.

George M. Beard, Griesinger's contemporary in the United States, wrote in 1869 that neurasthenia was caused by "dephosphorization of the nervous system;" M. Krishaber (1873) defined anxiety states as a cerebro-cardiac neuropathis; and so on. Practically all psychiatrists shared the opinion that mental disorders were diseases of the nervous system, either of hereditary-degenerative nature or caused by physicochemical factors. Some disorders were called "functional" in order to indicate that no organic factors have, so far, been discovered. Even Charcot, who practiced hypnosis, believed that neuroses were organic disorders of the nervous system.

When Freud studied medicine and was a budding physician, all mental disorders were called "nervous diseases," and disturbed individuals were believed to be "nervous." The few physicians who believed in the nonorganic origins of mental disorders were called "Psychiker," which was not a term of respect.

Freud studied medicine and later conducted research under Ernest Brücke (1819–1892) and Theodor Meynert (1833–1892). Brücke was a staunch supporter of reductionism. He believed that all psychological processes could be presented in physicochemical terms and all mental disorders were organic diseases. Meynert was a brain anatomist and an ardent follower of Griesinger. He adopted Herbart's mechanistic psychology and developed his own version of neurological associationism.

Scientific Psychology

In the last decades of the nineteenth century, psychology was dominated by Wilhelm Wundt (1832–1924) and his disciples. In the preface to the first edition of his *Principles of Physiological Psychology*, Wundt wrote in 1873: "The new discipline rests upon anatomical and physiological foundations."

One of Wundt's disciples, George Humphrey, described Wundt's historical role as follows:

> The effect of Wundt's work was roughly to change psychology from the concern of a number of nonexperimental, somewhat philosophically minded systematists, and a kind of side interest of a number of experimentalists in other subjects, into an experimental science.
>
> It is almost always the case that when one makes a statement about this man an immediate qualification comes from somebody. Someone may then say: but he was really a philosopher! Of course, he was. His book on *Logic* ran into four editions, as did that on *Ethics*. The writer

> was privileged to hear him lecture on the latter subject, as well as on others. His general treatise on philosophy ran into four editions, the last in 1919. But yet he was the Great Man of psychology, the international Shouter Down. On at least two continents his pronouncements provoked disagreement, and investigation by his own method, experiment. For, at times, although he shouted he did not by any means silence every opponent. And of course, an experimenter who provokes experiment is performing his own service, which is what Wundt did. (Humphrey 1968:276)

The other notable part of Wundt's experimental method was *analysis*. For, once again, keeping in the intellectual fashion, psychology must have its elementary data out of which the complex, shifting mass of experience can be built, and into which it can be analyzed. The psychological element is of course the sensation, which has its own attributes, such as intensity and clearness. Sensation is not an isolable fact of psychic life, any more than is the atom of physical matter but "a necessary assumption following the necessities of psychic analysis" (Wundt 1893 1:280), though, he thought, most chemical elements can be isolated.

Wundt maintained that *Elebnisse* (experiences) are the subject matter of psychology. These self-evident *innerlich gegeben* (givens from within) are observable data, and they can be observed from within, by the experiencing person. Freud had to raise the question as to these inner experiences could indeed be observed by the person who experiences them.

Thus, while the Zeitgeist was against Wundt introspection as evidenced by Pavlov's conditioning and Watson's behaviorism, Freud went in the opposite direction and turned the searchlight on unobservable, unconscious phenomena.

To make psychology an acceptable science, Wundt applied methodological reductionism and suggested combining introspection with experimentation. Wundt believed that psychological experimentation should follow the rules of natural sciences. Experiments create external conditions that look toward an objective and unbiased production of a determinate mental process. Experiments can isolate the relevant variables at will and make the scientific observer the master of the observed situation. Instead of naturalistic observation, experiments introduce rigorous control in the data to be observed (Wolman 1981: ch. 15).

According to Wundt, mind is not an object. Mind, as a subject matter of scientific inquiry, is a series of processes that take place in accordance with the causal laws of neurophysiology.

The elements of this conscious process are ideas, feelings, and impulses. The first comes from outside, the second and third from the organism itself. Ideas are composed of sensations. Sensations can be distinguished according to their quality and intensity. They are conducted by afferent nerves. Each sensation is followed by a muscular movement controlled by an efferent nerve. Association is a connection between sensation and movement and not, as in the old associationism, a connection between ideas.

Feelings do not come from any sense organ. They can be divided into pleasant or unpleasant, tense or relaxed, excited or depressed. Feeling is the "mark of reaction of apperception upon sensory content; feeling is active."

Although Wundt rejected the old mental faculties theory, he still divided all mental processes into reason, feelings, and volition. Volition can hardly be distinguished from feeling: Will is some sort of feeling, namely, a decision or resolution feeling, that leads to overt action. Volition is the very essence of life; it represents the needs of the organism and its tendency to purposive behavior.

Wundt maintained that psychology studies the processes of sensation, volition, and feeling. The unity of all these processes is called consciousness. The processes take place, change, and pass away in one's consciousness. What one can observe is the momentary state of consciousness or the "actuality" of it.

The unifying factor in mental processes is *apperception.* Wundt used the term apperception to denote assimilation, inclusion of new sensations, and their synthesis in the totality of consciousness. Feeling is the reaction of apperception to new sensory content. In Wundt's system volition implies active responses. Apperception is the focal point of the consciousness and the crucial element.

Wundt's work was continued by Titchener. "The primary aim of the experimental psychologist has been to analyze the structure of mind, to ravel out the elemental processes from the tangle of consciousness," Titchener wrote; psychology is a morphological science.

> I do not think that anyone who has followed the course of the experimental method, in its application to the higher processes and states of mind, can doubt that the main interest throughout has lain in morphological analysis, rather than in ascertainment of function. . . . The experimental psychology arose by way of reaction against the faculty psychology of the last century. This was a metaphysical, not a scientific, psychology. There is in reality, a great difference between, say, memory, regarded as a function of the psychophysical organism and memory regarded as a faculty of the substantial mind. (Titchener 1898:458)

Could Freud rebel against the prevailing "scientific" climate of his times?

Conflicting Influences

Psychoanalysis is a complex phenomenon. It was never meant to be a monolithic statue nor to be cast in one metal. Freud was influenced by conflicting cultural influences, such as the Judeo-Christian versus atheism, *Kulturwissenschaft* versus natural sciences, materialistic monism and positivism versus the German philosophical idealism, the organically oriented medical science which viewed mental disorders as diseases versus romanticism, hypnotism, and occult phenomena.

Freud was born in 1856 and died in 1939. The beginning of his scholarly and scientific career was parallel to the great intellectual tradition of the nineteenth century. His contemporaries were Alfred Russel Wallace, who lived from 1823 to 1913, Thomas H. Huxley (1825–1895), Thomas Edison (1847–1931), and Guglielmo Marconi (1874–1931). He was influenced by Charles Darwin (1809–1892), Gregor Mendel (1822–1889), and to some extent by Louis Pasteur (1822–1895) and Johann Friedrich Herbart (1776–1841).

But Freud was influenced not only by his contemporary scientific climate but also by the intellectual currents, literature, and art. Freud read not only Darwin's *The Descent of Man* (published in 1871), but he also read Nietzsche's *The Birth of Tragedy* (1872). He was as much influenced by Helmholtz and Brücke as by the works of writers like Emile Zola, Guy de Maupassant, Daudet, Maeterlinck, and Schopenhauer and by European mysticism. Freud's main studies were in the natural sciences and medicine, but he couldn't escape the influences of Renan, whose book on intellectual reform and morality in 1871 had an impact on all the culture of his times. Freud was also influenced by the plays of Henrik Ibsen (1828–1906) and Strindberg (1849–1912). Perhaps these writers, more than the scientists, anticipated the coming of Nazism and the building crisis of World War II. In the play, *Miss Julie*, which was published in 1888, Strindberg pointed out the growing decadence of human life and the conflict between intellectual aspects of life and human impulses. Freud was somewhat less influenced by Russian literature, but he was most impressed by Dostoyevsky, whose unusual insight into human life and human nature opened the eyes of many contemporaries, among them Freud.

Dilthey

Moreover, the highly *wissenschaflich* objective and experimental psychology was by no means generally accepted. Most psychologists abandoned Wundt's introspection and, under the leadership of Pavlov and

Watson turned to objective experimental research, but Wilhelm Dilthey, Spranger, Vaihinger, and others who exercised profound influence on Dilthey complained (1883) that

> . . . the contemporary experimental and physiologically oriented psychology applied the wrong method. Even natural scientists have found, under the influence of positivism, that science should be descriptive. Science should not go beyond observable data. . . .
>
> Experimental psychology deals with sensations and their associations and fails to see the man as he sees and feels himself. The higher mental processes are overlooked. Explanatory psychology is unable to see what poetry or autobiography or art or religion have seen. Art and literature cannot take the place of a scientific psychology, but explanatory psychology falls short of what it should be. There is no need for explanatory psychology, said Dilthey, and in 1894 explained his views as follows:
>
> "We know natural objects from without through our senses. However we may break them up or divide them, we never reach their ultimate elements in this way. We supply such elements by an amplification of experience. Again, the senses, regarded from the point of view of their purely physiological function, never give us the unity of the object. This exists for us only through a synthesis of the sense-stimuli which arises from within. . . . How different is the way in which mental life is given to us! In contrast to external perception, inner perception rests upon an awareness (Innewerden), a lived experience (Erleben), it is immediately given. Here, in sensation or in the feeling of pleasure accompanying it, something simple and indivisible is given to us." (Hodges 1949:133)

Dilthey has influenced some of Freud's associates, especially Alfred Adler and to a lesser extent, Otto Rank. Freud has offered another solution to the impasse pointed to by Dilthey. Freud asked the question, What is the motivating force? What is going on under the observable surface? Why do people do what they do?

Pavlov, Watson, the learning theorists, and the behaviorists broke away from Wundt and introduced more precise methods for the study of overt behavior. Spranger, Stern, and Gestalt psychologists broadened the cultural aspects of psychology. Freud and psychoanalyists reached beneath the surface and developed a "depth psychology." Freud turned towards the irrational elements of human nature, such as dreams, amnesias, and psychopathological symptoms. He discovered repression, projection, denial, and other defenses which made introspection a hopelessly ineffectual method. Freud's main discovery was the unconscious.

The Unconscious in Historical Perspective

Freud was not the first one to be concerned with the nature of unconscious processes.

René Descartes (1596–1650) decided to devote his life to the pursuit of truth. This decision put him in a state of growing excitement which "so exhausted him that the fire went to his brain and he fell into a kind of enthusiasm which so mastered his already cast down mind that it prepared it to receive the impressions of dreams and visions. . . . On November 10, 1619, having gone to bed *wholly filled with this enthusiasm* and wholly occupied with the thought that he had discovered that day *the foundations of the admirable science* he had three consecutive dreams in a single night, that he could only imagine had come to him from above," Descartes wrote about himself.

The first two dreams were frightening scenes of winds, rains, and storms. Descartes interpreted them as representing his faults and sins. The third dream offered a solution to the first two.

According to L. L. Whyte (1960:66 ff.) the terms *Unbewusstes* and *Bewusstlos* in German were first used in 1776 by E. Platner and appeared in the writings of Goethe, Schiller, and Schelling, the German writers of the Romantic era. In English language the term unconscious appeared in 1751, and was used frequently after 1800 in the writings of Coleridge and Wordsworth. In French it appeared much later, after 1850, translated from the German, and a French dictionary published in 1862 included the term *inconscient* and remarked that it was a rarely used word.

Gottfried Leibniz (1646–1716) introduced the idea of cognitive threshold and linked the unconscious to perceptual processes. According to Leibniz, "our clear concepts are like islands which arise above the ocean of obscure ones." However, the conscious concepts are merely islands or tips of a mountain, for most of our mental life is unconscious. Leibniz wrote that "it is not easy to conceive that a thing can think and not be conscious that it thinks," and yet this seems to be the truth.

Leibniz's ideas influenced J. F. Herbart (1776–1841). In Herbart's system, the discharges of mental energy are called "ideas." Since energy cannot perish, no idea can completely disappear. An idea inhibited by another idea sinks below the "threshold of consciousness." This process is called *repression (Verdrängung)*. The reappearance of repressed ideas meets with *resistance (Wiederstand)*. Apparently, Freud was considerably influenced by Herbart's theories (Jones 1953 1:374; Wolman 1968:42 ff.).

Both Kant and Herbart were much indebted to C. von Wolff (1679–1752). Wolff rejected the Cartesian dichotomy of conscious vs. unconscious and wrote (1725) as follows:

> Let no one imagine that I would join the Cartesians in asserting that nothing can be in the mind of which it is not aware. . . . That is a prejudice which impedes the understanding of the mind. . . .
>
> We conclude that we only then become conscious of objects when we distinguish them from one another; when we do not notice the difference of things which are presented to us, then we are not aware of what enters our senses. (Whyte 1960:102)

Immanuel Kant (1724–1804), despite his extreme rationalistic position concerning the human mind, accepted the concept of unconsciousness. He wrote:

> The field of our sense perceptions and sensations of which we are not conscious, though we undoubtedly can infer that we possess them, that is, the dark ideas in Man (and also in animals) is immeasurable. The clear ones in contrast cover infinitely few points which lie open to consciousness; that, in fact, on the great map of our spirit only a few points are illuminated; this can lead us to marvel regarding our own nature. (Wolman 1968:237)

Freud's Contemporaries

The developmental concept of unconsciousness was introduced by F. E. Benecke (1798–1854). He wrote that the mental life of the neurotic is entirely unconscious; he admitted there are alternate states of consciousness and unconsciousness. These unconscious powers possess from the very first an inherent capacity of becoming conscious. Consciousness, therefore, does not exist from the first in any soul; it must come into existence gradually.

Freud's theories carry close affinity to the works of the German philosopher Arthur Schopenhauer (1788–1860). In 1819 Schopenhauer published his main work, *Die Welt als Wille and Vorstellung* (The World as Will and Representation). According to Schopenhauer, man is driven by two unconscious forces, the instinct of self-preservation and the sexual instinct. Man is hardly aware of the fact that what he believes to be his own volition is actually dictated by these irrational and unconscious forces; he deludes himself believing that he makes his own decision in sexual behavior, while actually "fulfilling the demand of the species."

The sexual instinct is the "most important concern of Man and animal," and no other motive can compete with the sexual urge. Man is driven by the desire to copulate; his origin is in copulation and to copulate is "the wish of his wishes" (Schopenhauer 1819 2:457 ff).

Thomas Mann (1937) pointed to the similarities between Schopenhauer and Freud. Both men stressed the role of unconscious motivation and sexuality, and both views resulted in a rather grim perspective.

Another approach to the concept of unconscious was introduced in 1869 by Eduard von Hartmann (1842–1906) in his book *Philosophie des Unbewussten* (Philosophy of the Unconscious). Von Hartmann distinguished three states of the unconscious, namely, the substitution of absolute unconscious, the physiological unconscious, and the psychological or relative unconscious which controls most of human behavior.

Friedrich Nietzsche (1844–1890) stressed the importance of the unconscious. He

> conceived the unconscious as an area of confused thoughts, emotions, and instincts and at the same time as an area of reenactment of past stages of the individual and of the species. The obscurity, disorder, and incoherence of our representations in dreams recall the condition of the human mind in its early stages. The hallucinations of dreams also remind us of those collective hallucinations that seized while communities of primitive men (Ellenberger 1970:273)

Apparently Nietzsche influenced both Freud and Jung, but the latter borrowed from Nietzsche the idea of collective unconscious. According to Nietzsche, "In sleep and dream we repeat once more the task performed by early mankind" (1906 3:27).

Freud's contemporary, Henri Bergson (1859–1941), wrote as follows:

> To explore the most sacred depths of the unconscious, to labor on what I have just called the subsoil of consciousness, that will be the principal task of psychology in the century which is opening. I do not doubt that wonderful discoveries await it there, as important perhaps as have been in the preceeding centuries the discoveries of the physical and natural sciences. (Bergson, *The Independent,* October 30, 1913)

Freud did not discover the unconscious, but his was the first comprehensive study of unconscious processes. Freud adduced clinical evidence and offered a complex theory which linked childhood experiences with psychopathology and motivation and perception. Freud delved into the irrationality of human behavior and broke away from the rationalist tradition of Enlightenment and the physiologically oriented *Wissenschaft.* While he dealt with irrational process, he applied rational, scientific methods in their study. Freud's therapeutic aim was to transform the unconscious and irrational into conscious and rational.

Jung went further. He embraced irrationality and rejected the scientific principles of the Western Civilization. He built bridges leading to the Oriental philosophies of Zen and Tao, alchemy and mythology. His therapeutic method sought to establish a harmony between the conscious and the unconscious.

The theories described in the following chapters of this book greatly differ from one another. But all of them deal with the unconscious mind.

BIBLIOGRAPHY

Baruk, H. 1968. Nouvelles recherches sur Freud et la psychoanalyse. *Annales Médico-Psychologiques.* 1(4):595–601.

Bernfeld, S. 1944. Freud's earliest theories and the school of Helmholtz. *Psychoanalytic Quarterly*, 13:341–362.

Bernfeld, S. 1949. Freud's scientific beginnings. *American Imago*, 6:163–196.

—— 1949. Freud's scientific beginnings. *Imago*, 4:3–19; 6:188–204.

Boring, E. G. 1950. *A History of Experimental Psychology*, 2d ed. New York: Appleton-Century-Crofts.

Braun, O. 1909. *Eduard von Hartmann*. Stuttgart: Frommanns Verlag.

Brett, G. S. 1953. *History of Psychology*. R. S. Peters, ed. London: Allen and Unwin.

Burn, R. 1936. Sigmund Freud's Leistungen auf dem Gebiete der organischen Neurologie. *Schweiker Archiven für Neurologie und Psychiatrie*, 37:200–207.

Bykov, K. M. 1957. *The Cerebral Cortex and the Internal Organs*. New York: Chemical Publishing Company.

Cranefield, P. F. 1957. The organic physics of 1847 and the biophysics of today. *Journal of the History of Medicine*, 12:407–423.

Darwin, F., ed. 1892. *Charles Darwin: His Life Told in an Autobiographical Chapter and in a Selected Series of His Published Letters*. London.

Decker, H. S. 1977. Freud in Germany: Revolution and reaction in science, 1893–1907. *Psychological Issues*, 11 (Monograph 41).

Descartes, R. *Oeuvres*. Paris, 1908.

Dorer, M. 1932. *Historische Grundlagen der Psychoanalyse*. Leipzig: Fleix Meiner.

Du Bois-Reymond, E. H. 1918. *Jugendbriefe von Emil du Bois-Reymond an Eduard Hallmann*. Berlin: Reimer.

Dunlap, K. 1920. *Mysticism, Freudianism, and Scientific Psychology*. St. Louis: Mosby.

Ellenberger, H. F. 1956. Fechner and Freud. *Bulletin of Menninger Clinic*, 20:201–214.

—— 1970. *The Discovery of the Unconscious*. New York: Basic Books.

Fenichel, O. 1946. Some remarks on Fredu's place in the history of science. *Collected Papers,* 2:362–366. New York: Norton, 1954.

Freud, S. 1887–1902. *The Origins of Psychoanalysis: Letters to Wilhelm Fliess, Drafts and Notes.* New York: Basic Books, 1954.

—— 1891. *On Aphasia, a Critical Study.* New York: International Universities Press, 1953.

—— 1920. Beyond the Pleasure Principle. Standard Edition, 18:7–65. London: Hogarth Press, 1962.

Galdston, I. 1956. Freud and romantic medicine. *Bulletin of History of Medicine,* 30:489–507.

Gicklhorn, J. and R. Gicklhorn. 1960. *Sigmund Freud's akademische Laufbahn im Lichte der Dokumente.* Munich: Urban and Schwarzenberg.

Hartmann, E. von. 1869. *Philosophy of the Unconscious.* London: Kegan Paul, 1931.

Hoff, H. and F. Seitelberger. 1962. The history of the neurological school of Vienna. *Journal of Nervous Mental Disorders,* 116:495–505.

Humphrey, G. 1968. Wilhelm Wundt: The great master. In B. B. Wolman, ed., *Historical Roots of Contemporary Psychology,* pp. 248–274. New York: Harper and Row.

Jelliffe, S. E. 1931. Sigmund Freud as a neurologist: Some notes on his earlier neurobiological and clinical studies. *Journal of Nervous and Mental Diseases,* 85:696–711.

Jones, E. 1953–1957. *The Life and Work of Sigmund Freud,* 3 vols. New York: Basic Books.

Malthus, R. 1798. *An Essay on the Principle of Population as It Affects the Future Improvement of Society with Remarks on the Speculations of Mr. Godwin, M. Condorcet, and Other Writers.*

Mann, T. 1936. Freud and the future. In *Essays of Three Decades.* New York: Knopf, 1947.

Pribram, K. H. 1962. The neuropsychology of Sigmund Freud. In A. J. Bachrach, ed., *Experimental Foundations of Clinical Psychology.* New York.

Pribram, K. H. and M. M. Gill. 1976. *Freud's "Project" Reassessed: Preface to Contemporary Cognitive Theory and Neuropsychology.* New York: Basic Books.

Ritchie, B. F. 1965. Concerning the incurable vagueness in psychological theories. In B. B. Wolman and E. Nagel, eds., *Scientific Psychology.* pp. 150–165. New York: Basic Books.

Schick, A. 1964. The cultural background of Adler's and Freud's work. *American Journal of Psychotherapy,* 18:7–24.

Schopenhauer, A. (1819) *Die Welt als Wille und Vorstellung.*

Spehlmann, R. 1953. *Sigmund Freud's neurologische Schriften*. Berlin: Springer.
Taton, R., ed. 1965. *History of Science*, 3 vols. New York: Basic Books.
Tinterov, M. M. 1970. *Foundations of Hypnosis: From Mesmer to Freud*. Springfield, Ill.: Thomas.
Whyte, L. L. 1960. *The Unconscious Before Freud*. New York: Basic Books.
Wolman, B. B. 1968. The historical role of Johann Friedrich Herbart. In B. B. Wolman, ed., *Historical Roots of Contemporary Psychology*, pp. 29–46. New York: Harper and Row.
Wolman, B. B. Immanuel Kant and his impact on psychology. *Ibid.*, pp. 229–247.
Wolman, B. B. 1981. *Contemporary Theories and Systems in Psychology*. New York: Plenum.
Wundt, W. 1874. *Principles of Physiological Psychology*. London: Macmillan, 1940.

[PART II]
FREUD'S THEORY

[THREE]
First Principles

RATIONAL is defined in Webster as "having reason, agreeable to reason, sensible." Rationalism is defined as "the practice of guiding one's opinions and actions solely by what is considered reasonable." Reason is defined as "due exercise of the reasoning faculty; right thinking." Reasonable is defined as "just; fair-minded."

Western civilization is apparently committed to the belief in the rationality of human nature. Aristotelian logic has influenced our civilization to a much greater extent than has any other set of concepts. It has permeated Judaism and Christianity through the writings of Maimonides and Thomas Aquinas and has become the backbone of European thinking in modern times. Philosophers and scientists ascribed to and often prescribed logical roles to man and nature. It seems that thinking has become a partial revival of medieval scholastics in a secular disguise.

In Descartes' system the functions of reasoning and doubting have assumed crucial importance ahead of sensory impressions, and logical formulations have become the ultimate determinants of existence. Were Hamlet a Cartesian, his problem of being or not being would have been solved by logical inferences.

Descartes presented animals as mechanical creatures operating on some streamlined principles. Nature, as viewed by Descartes, was expected to follow the rules of formal logic (Eaton 1927).

This rationalist tradition came to its fullest development in Kant's critical idealism that subordinated man and nature to a priori forms and categories of the human mind. According to Kant (1881), the human mind, armed with the a priori forms of time, space, quality, quantity, causation, etc., became the sole judge of what was going on in the universe. Kant's disciple, Hegel, went even further, assuming that the entire world was ruled by the laws of dialectical logic in the sequence of thesis, antithesis, and synthesis. Hegel's *Absoluter Geist,* borrowed from Kant's transcendental mind, was to control human history and nature (Hegel 1902).

Empirical Psychology

The empirical revolution in modern psychology that started with Wundt's and Külpe's laboratories failed to break the rationalistic view of man, yet contemporary psychology professes to be an empirical science par excellence and refuses to be counted among humanities. In most universities, psychology belongs to the field of science and psychologists boast with their hairsplitting experimental designs. The methods of physical sciences have become the favorite path to be followed by psychologists as if in hope that physicalistic methods could render scientific rigor to psychology (Wolman 1965).

Yet, in all that unswerving determination and zeal for scientific and empiricist objectivity, contemporary psychology has managed to perpetuate the rationalist tradition, following Cartesian logic and Kantian reverence for mathematics. Whatever contemporary psychologists say has to be perfectly logical, impeccably quantified, and neatly fit into a set of concepts, arbitrarily set by the psychologists who wittingly or unwittingly idolize the mathematical order.

A good example of this rationalist tradition is offered by C. C. Hull's learning theory (1943, 1952). Whether one is a rat or a man, one is expected to behave in accordance with Hull's complex deductive mathematics. Quanta follow laws of probability, and living organisms are denied the right to rebel against them. The current discussion in the Hullian school does not affect less the fundamental principles of the theory than its inferential theories (Amsel 1965).

Tolman (1938) was less in awe of mathematical and logical inferences, but he paid the debt to rationalist tradition by ascribing "cognitive maps" and decision making to rats. Tolmanian rats must have been fairly rational beings to be able to follow the complex rules of their master. But, as the history of comparative psychology proves, experimental animals are not only rational but also docile. Thus, Thorndike's cat needed trial and error for learning, while Guthrie's cat learned at once (Wolman 1981).

Skinner (1953) paid his indebtedness to the rationalist tradition by excluding all subjective experiences from the realm of scientific inquiry. According to Skinner, science need not bother with "private" phenomena. Thus, "science" was purified from any subjective and potentially irrational elements. Skinner complained that Freud's constructs suggested dimensions to which the quantitative practices of science in general could not be applied (Skinner 1956:83).

Motivation

One should project Freud's contribution against this rationalist background. Freud's study of human nature was refreshingly free from past idiosyncracies. Perhaps one could describe *observable* actions in rationalistic terms or, at least, ascribe rationality to them, but Freud was intrigued not by *what* men do, but *why* they do what they do. The difference between Freud and most other students of human nature resembles the difference between chronicle and history. One can write down the manifest content of a dream, but it takes psychoanalysis to present factual data in meaningful order.

This is precisely Freud's contribution to psychology. Freud posed the "why" question, and this question led to the study of motivation. In his study of motivation, Freud had to fight both the reductionist tradition of medical sciences and the rationalist tradition of academic psychology.

One of the major psychoanalytic discoveries pertains to the variability of motivational factors. A hurt man may cry, scream, yell, or attack those who hurt him. A man may cry in pain, shame, anger, or even in joy. A man may desire a woman when he looks up to her or when he looks down at her, when she is practically inaccessible or when easily accessible, when she is smart and sophisticated, or naive and even stupid. Textbook wisdom and formal logic are of no help in these cases. A man may rationalize and make himself believe that he loves a woman *because* she is smart, but this may not necessarily be true. Rationalization may not at all correspond to truth, but merely reflect a desire to sound rational or be an unconscious maneuver to cover up the true and hidden motives. Men and women fall in love for a variety of reasons, usually not too rational nor logical. Rationalizations sound rational but are untrue, while the truth is not necessarily rational. Playwrights and novelists have often been more familiar with human feelings than academic psychologists, but why leave the study of motivation to artistic fancy?

Freud entered the field of motivation armed with the weapon of unrelentless determinism. Freud listened to people and analyzed the content of their communications. As crude as this method seems to be, the psychoanalytic confessions reflected a great deal of genuine human feelings as compared to the stylish elaborations of human nature produced by armchair philosophers or obtained in artificially set experiments or brought forward by nonhuman computers. Certainly Freud's research method lacked methodological sophistication and precision, but this simple mining method has dug out a lot of pure gold of human experiences while the sophisticated methods have sometimes produced no more than what was put into them by the experimenters. Freud's choice was between arid precision and less precise fruitfulness (Ross and Abrams 1965).

Ambivalence

Freud's empirical approach permitted the discovery of the ambivalences of human nature that could not be discerned by the rationalistic view. Love and hate, fear and courage, elation and depression sound like opposites to a logician. According to Aristotelian logic, opposites are mutually exclusive. Small wonder that logicians and rationalistic philosophers questioned whether one could hate and love the same person at the same time. Although the psychoanalytic descriptions of ambivalent feelings are a choice target for criticism, ambivalent feelings are a matter of fact, for what may not sound as a logical truth is nevertheless a psychological truth.

Freud made this choice on several occasions. Dreams do not follow the rules of Aristotelian logic but they reflect genuine human motivation. The study of dreams and of psychopathology of daily life led to the discovery of the unconscious and its irrational primary processes. Primary processes do not fit into textbooks of logic, and Freud's detailed description of illogical distortions and condensations in dreams and in symptom formation is a classic in research that may lack mathematical rules but does not lack discovery.

Thought Processes

Freud presented the problems of decision making in a complex model of personality. This model was *topographically* divided into a wholly irrational unconscious, a memory-filled preconscious, and a reality-oriented conscious. In addition, Freud's model was *structurally* divided into id, ego, and superego. The id is wholly irrational and operates on the principle of immediate gratification of needs (the so-called pleasure principle); the superego operates on moralistic grounds borrowed from environmental influences; and the ego, being somewhat more rational, makes decisions in anticipation of future consequences (the so-called reality principle). Obviously, Freudian *Homo sapiens* is far from being perfect, and not always is he very *sapiens*. Freud's *Homo sapiens* is merely a case in biological evolution; he certainly is not the epitome of Cartesian logic or the Kantian transcendental mind and comes much closer to the Darwinian concept.

Freud's theory of perception and reasoning is perhaps one of the weaker parts of the system (Arlow and Brenner 1964). This part of Freud's theory has never reached the clarity and depth that was attained in the structural theory. Freud was too much influenced by Herbart's mechanistic associationism and not too familiar with research in cognitive processes (Wolman 1964). However, the mainstream of Freud's theorizing leaves little doubt as to Freud's definite aversion to stylish Cartesian models of thinking.

The psychology of thought processes was under the influence of Descartes and Kant. Too often psychologists confused the issue of "how people think" with "how people *should* think" (Dewey 1910). The first issue belongs to psychology, the latter to logic. In a way, Freud's contribution paved the road for Piaget's research into how reasoning develops, and what kind of developmental stages the child goes through before reaching the more refined levels of abstraction, reasoning, and conceptualization (Piaget 1950, 1961).

It is, perhaps, a truism to say that logic does not describe the way most people reason, but merely sets the rules for proper reasoning. Yet this is the difference between psychology of thought processes and the rules of formal logic.

It was Freud's great achievement to break down the artificial barrier between normal and abnormal personality. Normality and abnormality are merely *degrees* of irrationality. Mentally disturbed individuals, having partially or completely lost control of their ego, are apparently less rational. No one is completely rational, and psychoanalytic treatment does not aim at converting patients into Cartesian or Kantian models. Psychoanalysis as a technique merely tries to improve one's personality balance and help the patients to reach an optimum of rationality.

Rational Science

Psychoanalysis as a science uses the same logical principles as all other sciences do. Because Freud laid bare the irrationality of human nature, his own writings have often been accused of irrationality. Science of the irrational may sound like a self-contradictory concept. "How," many asked, "could a science deal with irrational phenomena?"

The term science is used to denote both the activity of the scientist, that is scientific research, as well as the outcome of the research presented in a system of propositions. The aim of research is the discovery of truth, and the aim of scientific systems is the presentation and communication of the discovered truth.

The rationality of a science is entirely and totally dependent upon its methods of research, conceptualization, and communication. The rationality of science is not related to its subject matter, topic, and content. Consider mathematics, this undoubtedly rational and formal discipline. There is nothing rational about the square root of -1 ($\sqrt{-1}$), called by mathematicians *i*, which is an abbreviated sign for "irrational numbers." One can hardly find much rationality in the aggressive inter- and intraspecific behavior of fish and birds; however, the irrationality of animal behavior does not turn biology into an irrational discipline.

Life is not necessarily rational, and it does not subscribe to laws of logic, Aristotelian, Hegelian, or any other. It is quite easy to invent nonexisting rules and ascribe them to the phenomena under investigation. History of science knows of scores of such errors, and often the eradication of past errors counts among the most significant steps in the progress of scientific inquiry. One such error is the belief in man's innate rationality. Certainly science is a rational procedure, but the history of great discoveries and inventons does not prove that all progress was a product of carefully planned research designs. The fear of tackling the irrational, the unusual, and the nonquantifiable had a paralyzing effect on a great many scholars who forgot that rationality of a science is tested not by its subject matter, but by its truthfulness, by the logic of its inferences, and consistency of its communicated propositions.

Freud's Method

Some behavioral scientists seem to believe that all scientific propositions must be validated by controlled experimentation. However, there is no reason to dismiss the importance of clinical evidence. Psychoanalytic case studies are neither quantitative nor controlled, yet they bring a wealth of empirical findings and can be conclusive. As a rule the case study method should be the first step in inquiry. Case studies play a crucial role in finding out relevant variables. The necessity for this firsthand acquaintance with empirical data must never be overlooked in psychological research.

In some cases the prediction of an individual case carries tremendous weight as scientific evidence. Miss Mary H., twenty-eight years old, was avoiding male company and maintained that she never saw male sexual organs although she often took care of her young brother. Her dreams described a rapidly growing tree that burst out in fire and a few drops of water fell on her nude belly. The analyst suggested a physical checkup. To the surprise of the patient, the medical examination fully corroborated the analyst's suspicion. The patient was not a virgin; she had had sex relations at the age of ten. Further dreams brought in a sailor; her free associations led to a traveling salesman. It was her grandfather, a traveling salesman, who raped her. The clinical evidence of repression in this case seems to be conclusive (Wolman 1964).

Nagel (1959) wrote that Freud's theoretical propositions "were not proven." In fact, no theoretical propositions can be empirically proven.

As far as transcendent truth is concerned, psychoanalytic theory is, like any other theory, a set of constructs and hypotheses explaining empirical facts. It cannot be proven or disproven by direct empirical evidence.

A theory or a theoretical construct or model is scientifically useful when experimental or observational laws can be deduced from them. These experimental (in physics) and observational laws (in psychology) should be in a better agreement with observable data then any other laws.

Psychoanalytic theory is certainly free from logical contradictions even when it describes self-contradictory human behavior. It is logically true that the term love is the opposite of the term no-love, but it is empirically true that the very same individual may experience love and no-love toward the same person.

Freud's Postulates

Freud used the term "metapsychology" to describe his theoretical propositions (Meissner 1977). Rapaport and Gill (1959) have identified the following metapsychological principles:

1. The dynamic principle—that all behavior is driven by forces
2. The economic principle—that all behavior is a result of transformation of mental energies
3. The structural principle—describes the three parts of the mental apparatus
4. The genetic principle—stresses the origin of mental phenomena and their developmental stages
5. The adaptive principle—describes human behavior in terms of people's interaction with the environment, and adjustment for life.

The term metapsychology should be abandoned in favor of psychological theory of theoretical postulates. The following principles form the theoretical backbone of Freud's work:

1. Epistemological realism
2. Monism
3. Energetism
4. Determinism
5. Principle of economy
6. Pleasure-unpleasure principle
7. Principle of constancy

Epistemological Realism

Freud was definitely opposed to the logical positivism of the Viennese circle of philosophers. In the *New Lectures* (1932–1933), he wrote that scientific thought:

> carefully examines the trustworthiness of the sense perceptions on which it bases its conclusions, it provides itself with new perceptions which are not obtainable by everyday means, and isolates the determinants of these new experiences by purposely varied experimentation. Its aim is to arrive at correspondence with reality; that is, to toy with what exists outside us and independently of us. . . . This correspondence with the real external world we call "truth." (1932:233)

In contradistinction to logical positivism, epistemological realism demand both imminent and transcendent truth. Here is what Freud wrote about the Viennese circle philosophers:

> No doubt there have been intellectual nihilists of this kind before, but at the present time, the theory of relativity of modern physics seems to have gone to their heads. It is time that they start out from science, but they succeed in forcing it to cut the ground from under its own feet, to commit suicide, as it were. . . . According to this anarchistic doctrine, there is no such thing as truth, no assured knowledge of the external world. . . . Ultimately we find only what we need to find and see only what we desire to see. . . . And since the criterion of truth, correspondence with an external world, disappears, it is absolutely immaterial what views we accept. All of them are equally true and false. (p. 240)

Freud developed his theoretical framework on the grounds of epistemological realism. While radical empiricists, neopositivists, and operationists insisted that theory be merely a superstructure of sensory data, Darwin, Huxley, Sechenov, Pavlov, Brücke, Freud, Einstein, and other empirical scientists neither accepted the subjective sensory data as the sole basis for science, nor insisted on presenting a theory in terms of sensory data. The opposition to Kant, Mach, and logical-positivism crosses interdisciplinary lines.

Einstein advocated the idea, earlier proposed by H. Poincaré that "fundamental concepts and postulates" in science are "free inventions of the human mind." One must distinguish between facts and their theoretical interpretation. Theoretical interpretation is not a mere generalization or abstraction from facts. "Every attempt at a logical deduction of the basic concepts and postulates of mechanics from elementary experiences is doomed to failure," wrote Einstein in 1934 (p. 15).

Freud preached and practiced the principles of genuine empiricism, in the spirit of F. Bacon, J. S. Mill, and modern science. Whenever clinical data contradicted theory, the theory and not the data had to be changed.

In the *Outline of Psychoanalysis,* Freud wrote:

> We have adopted the hypothesis of a psychical apparatus, extended in space, appropriately constructed, developed by the exigencies of life, which gives rise to the phenomena of consciousness only at one particular point and under certain conditions. This hypothesis has put us in a position to establish psychology upon foundations similar to those of any other science, such as physics. In our science the problem is the same as in the others; behind the attributes (i.e., qualities) of the object under investigation which are directly given to our perception, we have to discover something which is more independent of the particular receptive capacities of our sense organs and which approximates more closely to what may be supposed to be the real state of things. (1938:105)

In the same outline, Freud wrote:

> Every science is based upon observations and experiences arrived at through the medium of our physical apparatus. But since our science has as its subject that apparatus itself, the analogy ends here. We make our observations through the medium of the same perceptual apparatus, precisely by the help of the breaks in the series of (conscious) mental events, since we fill the omissions by plausable inferences and translate them into conscious material. In this way we construct, as it were, series of conscious events complementary to the unconscious mental processes. The relative certainty of our mental science rests upon the binding force of the inferences. (p. 37)

Anxiety is a state of apprehension and tension is an empirical datum. Acts of love and hate are empirical data. A theory is not a mere generalization of data, but their interpretation. An interpretation is always a hypothetical statement that must be self-consistent, correspond to reality, and be superior to similar interpretations.

The correspondence to reality (transcendent truth) is the crucial test for any theory. Each field of science must establish its own *rules of correspondence* (Nagel 1961).

Monism

Freud was always aware of the organic foundations of mental life. At best, Freud believed, one can assume—and never do more than assume—that mental processes utilize a form of energy that is at the disposal of the living organism. This energy is analogous to any other energy, and that is all we know. "We assume, as the other natural sciences have taught us to expect, that in mental life some kind of energy is at work; but we have no data which enable us to come nearer to a knowledge of it by analogy with other forms of energy" (1938:44).

Being a monist, Freud never gave up hope for a monistic interpretation that would combine both physical and mental processes in one continuum. But, at the present state of scientific inquiry, a radical reductionism must be rejected. Psychology must continue to do what Freud actually did: develop new hypothetical constructs independent of the physical sciences. Freud knew that his later theoretical constructs were nonreductionistic, and irreducible to any of the constructs of physics or chemistry. Although he believed that the future might prove that chemical substances influence the amount of energy and its distribution in the human mind, work on such an assumption would not be too productive at the present time.

Quite late in his life Freud arrived at the conclusion that psychology must develop its own conceptual system, since the processes with which psychology is concerned

> . . . are in themselves just as unknowable as those dealt with by the other sciences, by chemistry or physics, for example; but it is possible to establish the laws which those processes obey and follow over long and unbroken stretches, their mutual relations and interdependences. . . . This cannot be effected without framing fresh hypotheses and creating fresh concepts. (1938:36)

Energetism

Freud believed that there is one kind of energy in nature and that all observable actions are either produced by this energy or exist as its variations or transformations. If this holds true in physics, it holds true also in other sciences, such as chemistry, biology, and psychology. This must not be construed in a radical reductionistic vein, for human thoughts are not electrical processes and cannot be reduced to terms of amperes, watts, or volts. Mental processes cannot be reduced to anything that is not mental, but they develop from the same physical source as everything else in the world. Mental energy is energy in the physical meaning of the word, that is, something that can be transformed into another kind of energy in a manner analogous to the transformation of mechanical into electric energy in generators. Energy can be accumulated, preserved, discharged, dissipated, blocked; but it cannot cease to exist. The law of preservation of mental energy, its transformability, and its analogousness to physical energy is one of the guiding principles of psychoanalysis.

> Among the psychic functions there is something which should be differentiated (an amount of affect, a sum of excitation), something having all the attributes of a quantity—although we possess no means of measuring it—a something which is capable of increase, decrease, displacement, and discharge, and which extends itself over the memory-traces of an idea like an electric charge over the surface of the body. We can apply this hypothesis . . . in the same sense as the physicist employs the conception of a fluid electric current. (Freud 1894:61)

Freud postulated that psychic energy is not an entirely new or a completely different type of energy. Mental energy is a derivative of physical energy, though no one can really tell how the "mysterious leap" takes place, either from body to mind or vice versa.

Determinism

According to Freud, objective and verified observation is the sole source of knowledge. The results of these observations can be "intellectually manipulated" and put together into a system of generalizations and laws that form a system of propositions that explain empirical data.

One of these general principles is the principle of causation. Natural sciences, especially microcosmic physics, struggle with the difficulties arising from a strict application of the causal principle. No such difficulties have been encountered in any of the areas of scientific psychology. All students of psychology apply a more or less strict deterministic point of view. Freud preferred a rigorous determinism that accepts no causes without effects, no effects without causes.

Determinism cannot be proven; it must be postulated and corroborated by empirical research. Once determinism is postulated, it forces the research worker to continuous efforts in seeking causes and predicting outcomes. Every successful case serves as evidence that one is on the right track, encouraging further efforts that promise to bring additional evidence. Lack of success indicates that one has to check and double-check one's methods and look for additional data. Strict determinism helped Freud in the study of the most irrational areas of dreaming and symptom formation in neuroses. The principle of "whatever is, has its causes" forced Freud to give up the early theory of instincts in order to assume the existence of destructive instincts. Causal considerations also put him constantly on guard in searching for minute details that might have been partial determinants of mental health and of mental disorder.

The Economy Principle

Freud's theory follows the principle of preservation of energy, and this principle is applied to the mental energy. Mental energy can be transformed, released, or accumulated; but it can never disappear entirely. When a degree of energy is invested into something, this object becomes loaded or charged with a certain amount of mental energy in a manner analogous to that in which bodies become charged with electricity. This process of charging ideas of objects with mental energy was called by Freud cathexis, and objects in which mental energy was invested were cathected. Cathexis can be applied to external objects as well as to one's own organism.

Energy is transformable and displaceable. Mental processes are processes of mental energy economics, that is, quantitative processes of transformation, accumulation, investment, and discharge of mental energy. Some processes consume more energy, some less. When powerful instinctual drives mass for an immediate discharge of energy, a great amount of energy is needed for anticathexis. Individuals torn by inner conflicts cannot be very efficient because considerable amounts of their energy are being tied in inner struggle.

Mental economy depends on the comparative strength of the external stimuli, instinctual drives, and the inhibitory forces. Human behavior can be presented as a series of reflex arcs. A stimulus acts on the organisms causing a disequilibrium (perceived as tension), and the tension leads to an action, that is, to a discharge of some amount of energy. The discharge of energy restores the equilibrium and is experienced subjectively as relief and pleasure.

Between the tension and discharge of energy two contradictory types of forces step in, one facilitating the discharge of energy that brings relief, the other preventing or postponing this discharge. The forces that urge and facilitate discharge are called by Freud drives or instincts. The instincts, or instinctual drives, press for discharge of energy, for lowering the level of excitation, and reduction of the tension in the organism. Thus these forces help to restore the equilibrium. Since homeostasis, or the tendency to keep equilibrium, seems to be a general property of living matter, the instinctual drives must be basic, innate, and primary biological forces.

The Pleasure-Unpleasure Principle

Freud followed Fechner with the idea of pleasure and unpleasure as related to the mental economy of excitation. Freud quoted Fechner as follows:

> Insofar as conscious impulses always have some relation to pleasure or unpleasure, pleasure and unpleasure too can be regarded as having a psycho-physical relation to conditions of stability and instability. This provides a basis for a hypothesis (that) every psycho-physical movement crossing the threshold of consciousness is attended by pleasure in proportion as, beyond a certain limit, it approximates to complete stability, and is attended by unpleasure in proportion as, beyond a certain limit, it deviates from complete stability; while between the two limits, which may be described as qualitative thresholds of pleasure and unpleasure, there is a certain margin of aesthetic indifference. (1920:8)

The ideas of constancy and economy were derived from clinical observations of pleasure and unpleasure, though from a logical point of view, the pleasure-unpleasure continuum should follow the principle of constancy. The mental apparatus endeavors to keep the quantity of excitation low, and any stimulus that increases the stimulation is felt as unpleasant.

> We have decided to relate pleasure and unpleasure to the quantity of excitation that is present in the mind but is not in any way "bound"; and to relate them in such a manner that unpleasure corresponds to an increase in the quantity of excitation and pleasure to a diminution. *(Ibid.)*

The Principle of Constancy

The principle of constancy serves as the general framework of Freud's theory of motivation. It represents a tension-relief continuum and explains the compulsion to repeat first experience. This "repetition compulsion" is responsible, and manifests itself in several aspects of human life.

> The attributes of life were at some time evoked in inanimate matter by the action of a force (of) whose nature we can form no conception. . . . The tension which then arose in what had hitherto been an inanimate substance endeavored to equalize its potential. In this way the first instinct came into being: the instinct to return to the inanimate nature. It was still an easy matter at that time for a living substance to die. For a long time, perhaps, living substance was thus being constantly created afresh and easily dying. (Freud 1920:40)

Freud invoked the constancy principle also in regard to the sexual instinct. Since "science has little to tell us about the origin of sexuality," Freud reported a myth that "traces the origin of an instinct to a need to restore an earlier state of things" (pp. 57–58). In Plato's *Symposium*, Aristophanes tells that "Everything about these primeval men was double; they had four hands and four feet, two faces, two privy parts, and so on. Eventually Zeus decided to cut these men in two." After the division had been made, "the two parts of the man, each desiring his other half, came together, and then their arms about one another eager to grow into one" (p. 60).

Freud hypothesized that living matter was broken down into small particles "which have ever since endeavored to reunite through sexual instincts." Several biological processes may be interpreted in the light of this tendency to restore an earlier state of things.

BIBLIOGRAPHY

Amsel, A. 1965. On inductive versus deductive approaches in neo-Hullian behaviorism. In B. B. Wolman and E. Nagel, eds., *Scientific Psychology, Principles and Approaches*. New York: Basic Books.
Arlow, J. and C. Brenner. 1964. *Psychological Concepts and the Structural Theory*. New York: International Universities Press.
Bykov, K. M. 1957. *The Cerebral Cortex and the Inner Organs*. New York: Chemical Publishing.
Dewey, J. 1910. *How We Think*. Boston: Heath.
Eaton, R. M. 1927. *Descartes: Selections*. New York: Scribner.
Einstein, A. 1934. *The World As I See It*. New York: Friede.
—— 1959. Reply to criticisms. In P. A. Schilpp, ed., *Albert Einstein: Philosopher Scientist*. New York: Harper and Row.
Freud, S. Standard Edition. James Strachey, ed. 24 vols. London: Hogarth, 1962; New York: Macmillan, 1962.
—— 1920. *Beyond the Pleasure Principle*. Standard Edition, 8:7–64.
—— 1933. *New Introductory Lectures on Psychoanalysis*. Standard Edition, 22:3–184.
—— 1937. *Construction in Psychoanalysis*. Standard Edition, 23:257–269.
—— 1938. *An Outline of Psychoanalysis*. Standard Edition, 22:141–208.
Hegel, G. W. F. *Philosophy of History*. New York: Colonial Press, 1902.
Hull, C. L. 1943. *Principles of Behavior*. New York: Appleton-Century-Crofts.
—— 1952. *A Behavior System*. New Haven: Yale University Press.
Kant, I. 1881. *Critique of Pure Reason*. London: Macmillan.
Meissner, W. W. 1977. Psychoanalytic metapsychology. In B. B. Wolman, ed., *International Encyclopedia of Psychiatry, Psychology, Psychoanalysis, and Neurology*. New York: Aesculapius Publishers.
Nagel, E. 1959. Methodological issues in psychoanalytic theory. In S. Hook, ed., *Psychoanalysis, Scientific Method, and Philosophy*. New York: New York University Press.
—— 1961. *The Structure of Science*. New York: Harcourt, Brace.
Piaget, J. 1950. *Introduction a l'épistemologie genétique*, 3 vols. Paris: Presses Universitaires.

—— 1961. *Les mécanismes perceptives.* Paris: Presses Universitaires.
Rapaport, D. and M. Gill. 1959. The points of view and assumptions of metapsychology. *International Journal of Psycho-Analysis,* 40:153–162.
Ross, N. and S. Abrams. 1965. Fundamentals of psychoanalytic theory. In B. B. Wolman, ed., *Handbook of Clinical Psychology.* New York: McGraw-Hill.
Skinner, B. F. 1953. *Science and Human Behavior.* New York: Macmillan.
—— 1956. Critique of psychoanalytic concepts and theories. In H. Feigl and M. Scriven, eds., *Minnesota Studies in the Philosophy of Science,* vol. 1. Minneapolis: University of Minnesota Press.
Sulloway, F. J. 1983. *Freud: Biologist of the Mind.* New York: Basic Books.
Teplov, B. M. 1957. Objective method in psychology. In B. Simon, ed., *Psychology in the Soviet Union.* Stanford, Calif.: Stanford University Press.
Tolman, E. C. 1938. The determiners of behavior at a choice point. *Psychological Review,* 45:1–41.
Wolman, B. B. 1964. Evidence in psychoanalytic research. *Journal of the American Psychoanalytic Association,* 12:717–733.
—— 1965. Toward a science of psychological science. In B. B. Wolman and E. Nagel, eds., *Scientific Psychology, Principles, and Approaches.* New York: Basic Books.
—— 1967. Johann Friedrich Herbart. In B. B. Wolman, ed., *Historical Roots of Contemporary Psychology.* New York: Harper and Row.
—— 1981. *Contemporary Theories and Systems in Psychology.* New York: Plenum.

[FOUR]
Mental Strata

FREUD'S early ideas have been described in his "Project for Scientific Psychology" (1895). Freud did not publish the project. As years went on, he discarded most of his early ideas and incorporated some of them in his later writings (Sulloway 1983: ch. 4).

The project, published posthumously, reflects Freud's early neurophysiological ideas. The mental apparatus was conceived as sort of a bodily organ; the mind was presented in a reductionistic fashion as being identical with the brain and the brain was believed to be comprised of *neurons*. Freud hypothesized three types of neurons. The *phi-neurons* were believed to perform sensory and perceptual functions; the *psi-neurons* were in charge of memory; and finally, the *omega-neurons* controlled all mental functions, in close analogy to Freud's future concept of the ego.

The fundamental function of the mind was to keep a state of equilibrium. Freud believed that the "principle of neuronal inertia" was the fundamental law of the mind. An excessive excitation caused disequilibrium and pain, and a discharge of energy restored the inner balance and brought pleasure. Sleep limits motility and thus prevents discharge of energy. Dreams represent wish fulfillment, for they permit hallucinatory discharges. Neurotic symptoms are also wishful states. They are composites of cathectic memory traces, that is, large accumulations of undischarged energy, especially sexual ones, evoked by parental seductions. These memory traces are unconscious.

Hypnotism

The fact that people can be unaware of their experiences but could be made aware of them was discovered by hypnotism. Anton Mesmer was a rather controversial figure and when he tried to demonstrate his theories of "animal magnetism" and "magnetic fluid," official medical science censured him, and rightly so. A special committee appointed by the French Académie des Sciences and the Faculty of Medicine in Paris concluded that "the magnetic fluid could not be noticed by any of our senses, therefore, such fluid does not exist" (Zilboorg and Henry 1940:245). Mesmer spoke of a "magnetic fluid" that allegedly fills the entire universe. A proper balance of this fluid in the human organism was supposed to bring good health. Mesmer maintained that he was restoring this magnetic equilibrium by touching his patients.

Of course, no such form of magnetism exists, but a British surgeon, James Braid, accepted the idea of influencing people's minds in a manner patterned after Mesmer. This new method was called *hypnotism.* In 1864, the French physician Liebeault wrote a book about the influence of "moral factors" upon the human body and described his treatment methods based on hypnotism. In 1882, the neurologist Jean Martin Charcot reported to the French Académie des Sciences far-reaching changes induced in hysterical patients under hypnosis.

Freud became acquainted with the hypnotic treatment of hysterics at the Salpetrière clinic in 1885. Charcot produced and removed hysterical symptoms by hypnotic suggestion. Charcot believed that psychogenic disturbances are a by-product of organic factors possibly related to heredity.

More or less at the same time, Pierre Janet in France developed a new theory of mental disorder. Although Janet linked mental disorder to heredity and, like the Italian scientist, Cesare Lombroso, saw in mental disease some "degeneration," his theory dealt with a nonorganic "automatism" that he called "unconscious."

In 1889 Freud went to the Nancy School and witnessed experiments in hypnotism conducted by H. M. Bernheim. The hypnotized subjects experienced hallucinations, but could not recall them when they woke up. When Bernheim urged the subjects, assuring them that they could recall, the subjects actually recalled everything.

Bernheim's experiments convinced Freud that people may not "know" what they *actually* know. Certain parts of one's mental life are concealed and not accessible to the consciousness. Bernheim's work proved that people are not always aware of what they are doing and quite often they are capable of making unconscious decisions. The traditional theory of will and willpower was finally discredited, and the rationality of human behavior put under a question mark.

The Unconscious

The use of hypnotism marked the beginning of the study of the unconscious.

Freud refused to accept the idea that what is inaccessible to our means of observation does not exist. In his search into the unobservable areas, Freud followed in the footsteps of all great discoverers, such as Newton, Darwin, and Pasteur. The fact that we haven't had direct contact with something, and we don't know how to find it, does not necessarily mean that what is unknown to us does not exist.

Freud's main contribution to the idea of the unconscious relates to a *topographic* concept of the human mind, and to the evidence that he adduced. According to Freud, the unconscious is not merely a way people have of experiencing mental phenomena, but sort of a province or layer of the human mind. Freud suggested the division of the human mind into three distinct layers, namely the unconscious (Ucs), preconscious (Pcs), and conscious (Cs). Freud termed this conceptualization "depth-psychology" or "topography." The three systems, Ucs, Pcs, and Cs, were defined on the basis of two fundamental criteria: their relationship to one's awareness and the mode of energy organization and functioning.

> In the first phase, the mental act is unconscious and belongs to the system Ucs; if upon the scrutiny of the censorship it is rejected, it is not allowed to pass into the second phase; it is then said to be *repressed* and must remain unconscious. If, however, it passes this scrutiny, it enters upon the second phase and thenceforth belongs to the second system, which we will call the Cs. But the fact that it so belongs does not unequivocally determine its relation to consciousness. It is not yet conscious, but it is certainly *capable of entering consciousness*. . . . In consideration of this capacity to become conscious, we also call the system Cs the *preconscious*. (Freud 1915:163).

> Intensity of cathexis is mobile in a far greater degree in this [Ucs system] than in the other systems. By the process of displacement, one idea may surrender to another the whole volume of its cathexis of several other ideas. I have proposed to regard these two processes as distinguishing marks of the so-called *primary process* in the mind. In the system Pcs the secondary process holds sway. (Ibid.)

Infantile Unconscious

Freud assumed that the mental processes of newborn infants are totally unconscious, although neonates are capable of perception. Inner and outer stimuli can work in them the feelings of pain or pleasure. These sensations and responses are believed to leave memory traces, but they are stored in the unconscious. These memories of early childhood experiences are inaccessible except in dreams, in hypnosis, and in psychoanalytic treatment which brings the unconscious to the surface. Many childhood wishes, for example, are unrealizable, but they still survive in the unconscious. Of course, it was and is impossible to observe directly the deep and preverbal layers of the human mind, but Freud discovered an *indirect* access to the unconscious. He studied the observable phenomena of dreams, free association, slips of the tongue, and the symptoms of mental disorder, and he traced from these the hidden forces of the mental layer of the unconscious.

Freud believed, in accordance with the homeostatic principle, that the infant's mind avoids overstimulation. Falling asleep leads to a disconnecting of the outer world, and the infant wakes up only under the pressure of disturbing stimuli such as hunger, bowel or bladder discomforts, pain, loud noises, and so on. As soon as he is able to rid himself of the stimuli, he shuts out the outer world and falls asleep again. However, external stimuli do not always permit the perpetuation of the equilibrium of relief and sleep, and tension-provoking stimuli often force themselves upon the infant and make him perceive them. The infant is forced to cope with them and try to master them. This may well be the origin of the realistic perception of the outer world and is probably the way that the conscious gradually emerges out of the unconscious. In this way, tension and frustration lead to the development of the conscious perception of reality.

The development of the conscious seems to depend on the infant's ability to utilize his past memories. When these memories accumulate on the "surface" of the unconscious, as it were, they are presumably more accessible than the deeper layers of the unconscious, and therefore may periodically become conscious. The part of the unconscious that may easily become conscious Freud called the *preconscious*.

Pcs and Ucs

The unconscious mind, then, is divided into two parts, the preconscious (Pcs) and the unconscious proper (Ucs). Preconscious processes may become conscious either spontaneously or by a conscious effort. Conscious processes are actually only conscious for a short while and they can easily submerge into the preconscious and the unconscious. The preconscious may be defined as that which is *in* our mind but not right then *on* our mind. Freud stated:

> Thus there are two kinds of unconscious, which have not yet been distinguished by psychologists. Both of them are unconscious in the sense used by psychology; but in our sense one of them, which we term the Ucs, is also inadmissible to consciousness, while we term the other Pcs because its excitation—after observing certain rules, it is true, and perhaps only after passing a fresh censorship, though nonetheless without regard to the Ucs—are able to reach consciousness. (1900:614–615)

People are at most times aware of only a small fraction of their mental processes at a given time, and the content of consciousness is pretty much a selective process. What is unconscious may become preconscious and even eventually conscious, but only a part ever becomes conscious. Some impulses and perceptions may become preconscious or conscious for a while and then be pushed down into the unconscious. Freud originally ascribed to conscious and unconscious dynamic properties, but in the end he relegated all dynamics to instincts and the "mental agencies" of id, ego, and superego (to be discussed later). The terms "conscious," "preconscious," and "unconscious," therefore, do not indicate any dynamic forces in personality but, as Freud finally concluded, the "mental provinces." They are purely topographic concepts indicating the "depth" of the mental processes and their relative distance from the surface. What an individual is aware of, he is conscious of; what he is not aware of, but may become aware of at any time, is preconscious; what he cannot become aware of without definite effort or cannot become aware of at all is unconscious.

Repression and Resistance

Freud distinguished between the conscious rejection of an impulse, i.e., *suppression* and *repression*. According to Freud, in suppression the energy put at the disposal of the impulse that is seeking to convert itself into action has been withdrawn. The impulse has no more energy attached to it, and it becomes powerless, existing merely as a memory. Repression, on the other hand, is a vehement effort that has been exercised to prevent the mental process in question from penetrating into the conscious and as a result it has remained unconscious. Repression may be defined as an act by which either a mental act is prevented from entering into the preconscious and forced back into the unconscious or a mental system that belongs to the preconscious is thrown into the unconscious.

Whatever has been repressed constantly attempts to find outlets for discharge. Freud was always consistent in regard to the energetic principle, interpreting all mental processes as processes in which some energy is accumulated, stored, blocked, or discharged. Some forces, whether conscious or unconscious, prevent the discharge of energy. These repressing forces or "defense mechanisms" forever resist the discharge of repressed energy. Patients in the process of psychoanalytic treatment all act as if they object to being cured. Although they come for treatment voluntarily, expending time and money, they seem to avoid doing what they are requested to do. They may consciously cooperate with the therapist, but somehow they manage not to follow his instructions, as if opposing the unraveling of the unconscious conflict and the very process of cure. Very simply, the patients unconsciously "resist" the efforts of the therapist that are aimed at the discharge of blocked energies and the removal of tension. *Resistance,* then, is nothing but a continuation of repression, and it requires a special therapeutic skill to undo the resistance and to release the repressed energies.

Amnesia

Pavlov (1928) explained the fact of forgetting in terms of extinction; whatever has been conditioned may become unconditioned unless it is reinforced. Extinction cannot explain, however, the fact of complete loss of memory that includes well-known and otherwise well-remembered facts, and it cannot account for the strange cases of forgetting one's own name, address, and occupation in a shell-shock traumatic neurosis.

For Freud, the most intriguing aspect of amnesia was not the loss of memory, but its recovery. Freud was curious about where the hidden memories go and how they could reappear after a substantial lapse of time. Since memories of the past can come back, Freud reasoned, apparently they are not completely lost. Most adults have long forgotten their early childhood, but they can recall it under hypnosis and in psychoanalysis. Freud's answer to the question of where lost memories were hidden was simple: in the unconscious. The unconscious is not therefore an hypothesis, it is a fact. The recall of forgotten material shows that what is forgotten is not really lost, but is merely stored away, and Freud called this storage area the unconscious. In cases of "split personality," where the patient loses his identity and regains it, whether spontaneously or in hypnotic or psychoanalytic treatment, his dissociation is unconscious; his past experiences are buried there and are brought to the surface in the process of regaining self-awareness.

Often people seem to forget significant events and recall inconsequential childhood experiences. In the process of psychoanalytic treatment, Freud found that irrelevant memories substitute for some really significant impressions whose direct reproduction is prevented. He found that often unimportant events were recalled not because of their own contents, but because of their association with important but repressed thoughts. These he called "concealing memories." In some cases, these concealing memories were associated with the first years of childhood, while the thoughts represented by them that remained unconscious belonged to a later period of life. Freud called this phenomenon *retroactive* or *regressive displacement.* In other cases, an irrelevant impression from earliest childhood is associated with an even earlier experience that cannot be recalled. Freud called this type of concealing memory *encroaching* or *interposing memories.* Finally, concealing memories may be connected by simple temporal contiguity with the impression they are actually trying to conceal. This type of concealing memory is called *contemporaneous* or *contiguous.* In all of these cases, what is concealed is concealed in the layer of the unconscious.

Herbart's Influence

Apparently, Freud was influenced by J. F. Herbart and ascribed dynamic properties to both unconscious and conscious. The unconscious, *primary* energy is unbound, free-floating, and not goal directed. The conscious, *secondary* energy is bound and goal directed. Freud wrote as follows:

> All that I insist upon is the idea that the activity of the first system is directed towards securing the *free discharge* of the quantities of excitation, while the second system, by means of the cathexes emanating from it, succeeds in *inhibiting* this discharge and in transforming the cathexis into a quiescent one, no doubt with a simultaneous raising of its potential. . . . The second system succeeds in retaining the major part of its cathexes of energy in a state of quiescence and in employing only a small part on displacement. . . . When once the second system has concluded its exploratory thought-activity, it releases the inhibition and damming-up of the excitations and allows them to discharge themselves in movement. . . . The second system cathects memories in such a way that there is an inhibition of their discharge, including, therefore, an inhibition of discharge (comparable to that of a motor innervation) in

> the direction of the development of unpleasure. . . . It would be an unnecessary expenditure of energy if [the second system] sent out large quantities of cathexis along the various paths of thought and thus caused them to drain away to no useful purpose and diminish the quantity available for altering the external world. (1900:599–601)

In Herbart's system, perceptions *(Vorstellungen)* were the indivisible elements of mental life. These elements, however, were viewed by Herbart both as units of energy and as psychological concepts of perception. Thus, in Herbart's system, a "brighter"—a more precise—perception "pushes" itself up in the consciousness as if the perception were a physical dynamic unit.

Primary Processes

Something similar to Herbart's *Vorstellungen* was conveyed in Freud's early work. In his "Project for Scientific Psychology," primary processes indicate the urge for an immediate action, come what may. This urgency was called by Freud the "pleasure principle." Pleasure in this context is rather a poor translation of the German *Lust;* the pleasure principle should be called, rather, the principle of lust, of urge (cf. Freud 1887–1902; Holt 1962).

Freud made the need for an immediate gratification isomorphic, if not identical, with the need for a physical discharge of energy. Freud's primary processes are full of symbolization, condensation, displacement, and fusion in their *meaning*. At the same time, the primary, that is, the unconscious processes are characterized by a high tension of unbound energy that presses for immediate discharge. The conceptual irrational content of primary processes was fused by Freud with their dynamic property—their action parallels that of a bursting volcano.

The conscious, secondary processes are conceptually rational, goal directed, logical, and reality-oriented. Mature and normal individuals perceive things as they are and comprehend them in a rational way. All these are secondary processes. On the dynamic side, the secondary processes reflect quiescent energy, controlled action, and delay in the discharge of energy dictated by survival considerations. The ability of delaying, controlling, and anticipating future results was called by Freud the "reality principle." Both the pleasure and the reality principles will be explained in detail in connection with the id and ego theories.

In the later development of his psychological theory, Freud transferred the dynamic concepts of energy to libido and its destructive counterpart, located in the id; and the control apparatus was conceptualized as ego and superego. Some psychoanalytic thinkers question the necessity of the topographic "provinces" (Arlow and Brenner 1964; Gill 1963). Freud himself viewed the terms conscious, preconscious, and unconscious not as dynamic forces in personality, but, as the "mental provinces." They are "topographic" concepts indicating the "depth" of the mental processes (Holt 1976).

Dreams

Freud viewed sleep as a turning away from the outside world. He believed that man's relationship with the world, which he entered into so unwillingly at birth, was endurable only with intermission, so that he tries to withdraw again into the condition prior to his entrance into the world; that is, into intrauterine existence. (Some individuals even as adults roll themselves tightly up into a ball when sleeping, resembling the intrauterine position).

Freud stated that a dream

> . . . may be coherent, smoothly composed, like a literary work, or unintelligibly confused, almost like a delirium; it may have absurd elements, or jokes and apparently brilliant inferences; it may seem clear and well defined to the dreamer, or it may be dim and indefinite; the pictures in it may have the full sensuous force of a perception, or they may be as shadowy and vague as a mist. (1933:79)

It is possible to wake up from a dream full of hope and joy with the heart beating fast, expecting good things to come. Or it is possible to wake up in a confused and perplexed mood, full of doubts, anxieties, and worries. The sleeper may even be afraid to lift his head after a dream of crime, death, and horror, or he may wake up screaming about a Frankenstein monster. On the other hand, he may not wish to get up at all if his dream has made him feel guilty. In most cases, even highly unpleasant dreams do not awaken the sleeper, and in fact, dreams have been referred to as the guardians of sleep.

Dreams are the mind's reaction to external or internal stimuli that act upon the organism in sleep. When dreaming, the individual attributes objective reality to the imagery that forms the material of his dreams. Freud found that a dream holds all the absurdities, delusions, and illusions of a psychosis (Wolman 1979).

In Freudian terms, sleep may be defined as a willful and temporary withdrawal from the external world and a regression to the state of mind that preceded the development of the conscious. Dreaming represents a compromise between the inner or outer disturbance of sleep and the desire to sleep. The sleeper "dreams away" the disturbance and continues to sleep. According to Freud, dreams most often reflect wishes stemming from the unconscious and these wishes are repressed demands for instinctual gratification. These demands often contain residues of the day's activities in the waking state, and may be related to a decision that has to be made or to a conflict to be resolved.

The unconscious demands themselves may be described as pressures coming up against the conscious part of personality; the part that represses the unconscious wishes. In Freud's classic *The Interpretation of Dreams* (1900), he likened the conscious mind to a censor which prevents the unconscious from becoming conscious. "The censorship between the Ucs and the Pcs, the assumption of whose existence is positively forced upon us by dreams, deserves to be recognized and respected as the watchman of our mental health" (Freud 1900:567).

In 1912 Freud clarified some of his ideas on the nature of the unconscious. "The term *unconscious*, which was used in the purely descriptive sense before, now comes to imply something more. It designates not only latent ideas in general, but especially ideas with a certain dynamic character, ideas keeping apart from consciousness in spite of their intensity and activity" (1912:262). And further on: "Every psychical act begins as an unconscious one, and it may either remain so or go on developing into consciousness, according as it meets with resistance or not."

Censorship

As early as 1915, however, Freud realized that the censorship force is not necessarily conscious, and that the repressing force may be itself unconscious.

> The reason for all these difficulties is that the attribute of being conscious . . . is in no way suited to serve as a criterion for the differentiation of systems. . . . [Thus] not only the psychically repressed . . . remains alien to the consciousness, but also something . . . that forms the strongest functional antithesis to the repressed. (Freud 1915:192–193)

In 1900 Freud assumed that the unconscious was the dynamic force of hidden desires, while the conscious censored such wishes and worked to prevent them from coming true. A dream is seen as a compromise, and through the dreamwork, the unconscious forbidden wishes become disguised and as such, are allowed to appear as a manifest dream. The censoring force repressed the forbidden wish, but it came through in dreams, errors, and neurotic symptoms.

In a later stage of his work, Freud (1923) transferred the censorship function to the ego. In dreams the ego holds in check the forbidden wishes "by what appears to be an act of compliance: it meets the demand with what is in the circumstances the innocent fulfillment of a wish and thus disposes of the demand. This replacement of a demand by the fulfillment of a wish remains the essential function of dream work" (Freud 1938:55).

Freud called the unconscious wish the *latent dream thoughts* and he referred to the story reported by the dreamer (i.e., the dream itself) as the *manifest dream content.* Freud gives the example of a physician who was supposed to be in a hospital very early. The physician was tired and he dreamed that he was a patient in the hospital who did not have to get up early in the morning. This was the manifest dream, but the latent thought was the wish not to go to the hospital. The physician's sleep was protested by means of the *dreamwork* that transformed the unconscious wish into the manifest dream content of being already in the hospital as a patient.

It was clear to Freud, then, that dreams reflect hidden wishes that cannot be accepted in the conscious, waking state. Every dream represents for Freud an attempt to protect sleep and to put aside a disturbance of sleep by means of a wish fulfillment. Since the path to motor discharge is closed in sleep, the wish fulfillment of the repressed impulse is toned down, and the repressed impulse that presses toward motor discharge has to content itself with an hallucinatory satisfaction. The latent dream thoughts must be turned into a collection of sensory images and visual scenes, and this transformation of a wish into a dream is called *dreamwork* (Freud 1933).

Dreams as Wish Fulfillments

By analyzing his own dreams, Freud discovered that wish fulfillments are the essence of dreams. Freud presented the dream cited below as an example of one of his own dreams. At the time of the dream (1900), he had been treating a friend for somatic symptoms related to anxiety hysteria. Irma's treatment had been temporarily discontinued for the summer, as she was reluctant to accept one of Freud's interpretations. On the day preceding the night of the dream, Freud asked another friend,

Otto, how Irma was doing. "She is better, but not quite well," Otto had replied. In this reply Freud sensed a note of reproach, and the same evening, in an effort to justify himself, he prepared the case history of Irma to present to another physician, Dr. M. That night Freud reports that he had the following dream.

> A great hall—a number of guests, whom we are receiving—among them Irma, whom I immediately take aside, as though to answer her letter, and to reproach her for not accepting my solution. I say to her: "If you still have pains, it is really only your own fault." . . . She answers: "If you only knew what pains I have now in the throat, stomach, and abdomen—I am choked by them!" I am startled, and look at her. She looks pale and puffy. I think that after all I must be overlooking some organic affection. I take her to the window and look into her throat. She offers some resistance to this, like a woman who has a set of false teeth. I think, surely, she doesn't need them.—The mouth then opens wide, and I find a large white spot on the right, and elsewhere I see extensive grayish-white scabs adhering to curiously curled formations, which are evidently shaped like the turbinal bones of the nose.—I quickly call Dr. M., who repeats the examination and confirms it. . . . Dr. M. looks quite unlike his usual self; he is very pale, he limps, and his chin is clean-shaven. . . . Now my friend Otto, too, is standing beside her, and my friend Leopold percusses her covered chest, and says: "She has a dullness below, on the left," and also calls attention to an infiltrated portion of the skin on the left shoulder which I can feel in spite of the dress. . . . M. Says: "There's no doubt that it's an infection, but it doesn't matter, dysentery will follow and the poison will be eliminated." We know, too precisely, how the infection originated. My friend Otto, not long ago, gave her, when she was feeling unwell, an injection of a preparation of propyl . . . propyls . . . propionic acid . . . trimethylin (the formula of which I see before me, printed in heavy type). . . . One doesn't give such an injection so rashly. . . . Probably, too, the syringe was not clean. (1900:107)

Through the careful analysis of the "Irma's Injection" dream by free association, Freud found that the dream had actually fulfilled some of his own wishes. Freud was not responsible for Irma's pain in the dream because the pain was organic. In addition, Freud took revenge on Otto in the dream, accusing him of careless medical treatment (the unclean injection), and also comparing him with his colleague Leopold. The dream also allowed Freud to take revenge on Irma by making her look "pale and puffy." Dr. M., who also disagreed with Freud's interpretation of Irma's case, was made to appear ignorant in the dream. Freud thought later that in the dream Dr. M. had resembled Freud's brother. Freud had been on bad terms with his older brother, who had recently rejected his advice.

The "Irma's Injection" dream also demonstrates some aspects of the concepts of manifest dream content and latent dream thoughts. In some dreams the manifest element may be only a fragment or an illusion substituted for the total latent thought, while in others, the manifest dream may be not so much a distortion of the latent thoughts as a representation, using concrete imagery, of words only remotely related to the latent thoughts. Some dreams reflect abstract thoughts that are substituted in the manifest dream by images that conceal or even contradict the latent thoughts, while in other dreams the latent thoughts may appear unaltered, or even intensified, in the manifest dream.

Children's dreams seem to demonstrate less distortion than those of adults, probably because their impulse control is weaker. Freud's daughter had a dream after she had crossed the Aussee by boat. Apparently the trip was too quick an experience for her, because that night she dreamt that she went sailing on a lake. This dream, like children's dreams in general, demonstrates that dream thoughts are usually related to recent experiences, are meaningful, and fulfill a wish.

Dreams that are stimulated by physical needs are usually very poorly disguised and the wish fulfillment is often quite transparent. Freud himself could be assured of having a certain type of dream by consuming large quantities of salted food just before going to bed. He would invariably become thirsty while asleep, and just before awakening he would have a dream in which he was drinking water. He called these dreams "dreams of convenience" because they fulfill a wish by attempting to provide the dreamer's need without waking him up.

Dreamwork

Through the analysis of dreams, Freud was able to discover the laws that govern the "primary" unconscious processes and to find the differences between them and the "secondary" conscious processes. According to Freud, in primary processes the thought elements that are kept apart in the waking state are *condensated* into larger units. A single element of the manifest dream, for example, may stand for a whole conglomeration of unconscious, latent dream thoughts, and a short manifest dream content is usually an *allusion* to a variety of unconscious thoughts.

In the waking state, instinctual energies are bound and cathected in definite objects, but in dreams, those cathexes, or mental energy loads, drift and shift easily from one object to another. In some dreams significant issues are barely mentioned, or are represented by trivia, or perhaps are only alluded to in the manifest dream content, while unimportant unconscious elements may be represented in the manifest dream in a very clear manner. This process of the dreamwork Freud called *displacement.*

Nonrelated elements in dreams are put together and represented as a whole by the process of *secondary elaboration.* Often an idea is represented by its opposite, and a causal relation may be substituted for a coincidental sequence in time. In many cases the manifest dream goes contrary to the wish, as if saying, "If only things could be the other way around." Freud called this distortion *inversion.* Inversion occurs not only in content but in time sequence as well and in many words that have multiple, contradictory meanings (e.g., the Latin word *altus,* which means both high and deep.)

The dreamwork, of course, has its limitations. It cannot, for example, create conversation, for any conversation included in the dream is usually that which has been recently heard, and it cannot confer judgment or register surprise. Critical powers and the power of deductive reasoning are suspended in the dreamwork, as they are the province of the conscious mind.

By its very nature, the dreamwork takes the dreamer back to the primitive era of intellectual development in which the individual's early childhood and the childhood of mankind are represented. Freud explained this regressive aspect of dreams by stating that our perceptions leave *memory traces* in the unconscious mind. In the waking state, the censorship systems are too active to allow the preconscious to bring these repressed memories into consciousness. In sleep, however, the memories gain access to the conscious through a regression from a complex ideation into the primary, raw material of the underlying memory traces. Mental energy, since it has no outlet in any physical motion during sleep, takes an active part in pushing the unconscious thoughts toward the conscious, thus providing the impetus upon which the dreamwork makes the unconscious material palatable for the conscious.

Another characteristic of the dreamwork is that it often represents two contrary ideas by the same composite structure. Each element in the content of a dream appears to be *overdetermined* by material in the dream thoughts and it is not derived from a single element in the dream thoughts, but may be traced back to a complex aggregate of ideas not necessarily related to one another. Just as connections lead from each element of the dream to several other elements, so a single dream thought is usually represented by a variety of dream elements. Condensation, together with the transformation of thoughts into imaginary actions *(dramatization)*, is one of the most peculiar characteristics of the dream work.

Some dreams are so complicated and confused that condensation and dramatization alone are not enough to account for the whole story of the manifest dream. In these dreams, the above-mentioned phenomenon of displacement takes place, transforming the psychical intensity, the latent thoughts and wishes into sensory experiences.

Dream Thoughts

The dream thoughts often follow certain logic which becomes distorted by the dreamwork. Some dreams disregard the logical structure, while others indicate it. A dream may reflect the logical structure of the dream thoughts by combining the material into a unity. Logical sequences may be represented by simultaneity, and when a dream shows closely related elements, it also indicates that they are connected in the dream thoughts.

Causal relations are often represented in a dream as being comprised of two parts, a subordinate part and the principal part. The principal part is long and elaborate. Causal relations are sometimes represented by transformation of one image into another. The transformation is visible, pictorial, and it represents causal relationship as a matter of mere temporal succession.

Dreams cannot express alternatives, such as either-or, but convey them as if they were two equal alternatives. Quite often a dream represents any element by its opposite.

Dream formation represents the logical relations of similarity, agreement, and contiguity. Identification is used when unity is already found in the dream material, and especially when the dream is concerned with persons. Composition is used when the unity is being freshly created. Identification consists in representing in the manifest dream content one of two or more interrelated people. A composite person may appear in a dream, a result of a union of the features of the interrelated persons. Identification and composition serve to avoid representing the features common to two or more people, and thus to bypass the dream censor. Sometimes, the common feature is represented in the dream as a clue to another hidden feature. Identification and composition usually serve the purpose of representing a characteristic common to two people, or of a displaced common feature, and expressing a composite of features that is wished for.

Composite formations often cause weird elements of dreams. Composite formations represent the characteristics of one thing, which is referred to another object, and/or combine characteristics of one object with those of another to make a new image.

Inversion, or transformation into the opposite, enables the dream censor to distort the dream thoughts. Besides inversion of content, the dream censor frequently inverts the beginning of the dream with the end.

The intensity of dream images vary. Somatic sensations during sleep are less vivid as the mental impressions based on memories. The intensity of the manifest dream has little to do with the intensity of the elements in the latent dream, for the latent dream elements which require the most extensive condensation appear in a most intense manner in the manifest dream.

The opposite of distinctness in dream is haziness, and in the case of distinctness of an entire dream, the opposite of distinctness is confusion. Distinctness or indistinctness of a dream is produced by the dream thoughts. Contradictions and inner conflicts are usually represented in a dream by inhibited movement.

Dead people may appear in dreams as alive and active. Dreams in which the deceased is first living, then dead, and finally living, often represent the willful though unconscious indifference of the dreamer and his ambivalent feelings toward the dead person.

Some dreams are absurd, if the dreamer thinks that his dream is nonsense, and his unconscious thoughts represent criticism and ridicule. Thus, by absurdity, the dreamwork expresses contradiction. The dreamwork also may represent contradiction by inverting the material related between the dream thoughts and the dream content, and the feeling of motor inhibition. The absurdity of a dream repeats the dream thoughts' tendency to express ridicule together with contradiction. The dream thoughts as such are never absurd. For whatever takes place in a dream of the nature of intellectual criticism is part of the dream thoughts, and it is not an intellectual performance of the dreamwork. Judgments made upon the dream after a waking are a part of the latent content of the dream. An act of judgment in the manifest dream is a repetition of an original act of judgment in the latent dream content.

Emotional Elements

The ideational content of a dream is not necessarily followed by emotions. One may experience in his dream a dangerous or embarrassing situation, yet may feel neither fear nor aversion. In some dreams the emotional content is connected with the ideational content, while in others the emotional content is separated from its conceptual content and expressed elsewhere in the dream. Emotional aspects of a dream are usually conveyed in the dream thoughts, but the reverse is not true. A dream is usually not as rich in emotions as the material from which it is composed, for the dream censor represses emotions.

The opposition of the dream thoughts results in an indifferent emotional content. The dreamwork can distort the latent content of their emotions, reduce them, or transform them into their opposites. Every element of the dream can be represented by itself or by its opposite. Transforming an element into its opposite is displacement which serves the purpose of the dream censor but it may be also wish fulfillment, substituting an unpleasant thing by its opposite.

Sometimes, the ideational content of a dream determines the emotional content, and at other times the ideational content is produced by the emotional disposition. The dreamer's mood receives the same treatment as the emotion emerging during sleep, which is either disposed of or interpreted for the use of wish fulfillment. Painful moods awaken strong desires.

Daydreams are similar to actual dreams for both are wish fulfillments based upon childhood impressions. The secondary elaboration operates in the same manner in sleep and daydreams. Whether a daydream will be repeated in a dream depends on its advantages against the censor and its ability to be condensed. The dreamwork usually prefers to use the ready-made fantasy of a daydream, instead of creating one from the dream thoughts.

Forgetting Dreams

Usually one has more dreams than one can remember, and one cannot be certain how many stories he remembers. If one doubts as to whether he is indeed accurately reproducing his dream, this is caused by the dream censor, which resists the emergence of the dream thoughts. Psychoanalysis revives forgotten dreams. A suddenly remembered part of the dream is usually its most important part.

Dream Language

Since undisguised wishes could not get by the repression of the "censor," the dreamwork uses pictorial symbols to give disguised representation to the latent dream wishes. These symbols are residue of forgotten identity, and, in general, they can be seen as a reflection of an archaic mode of thinking. In fact, our knowledge of symbolism is based on cultural-historical sources such as mythology, fairy tales, folklore, folk songs, and so on. For example, in ancient Hebrew a woman is often described as a house; when she is no longer a virgin her "door" has been opened. The fetus lives in a bog of water and mammals are descended from water-inhabiting organisms, so dreaming of water is thought to symbolize birth. Dreams of travel often represent death, the last journey, and in general, little animals and vermin symbolize siblings and young children. Parents usually appear in dreams as kings and queens or other authority figures.

Freud found that many dream components relate to sexual organs and functions. Cupboards, boxes, carriages, ovens, flower pots and vases, receptacles, bags, purses, drawers, jars, doors, and gates are female sexual symbols. Entrances to rooms and stores symbolize female genitals. Apples, peaches, and oranges represent breasts, and the number two often stands for the two breasts. Landscapes with rocks, woods, and water have been found to represent female organs. Cloaks, sticks, and hats, on the other hand, symbolize men, and guns, swords, knives, poles, pencils, snakes, tree trunks, and hammers symbolize the male sexual organ. A bridge often represents the male organ that connects parents during sexual intercourse, and may also indicate birth or death (i.e., the transition from life to death or womb to birth). Airplanes and balloons symbolize erection, while flying symbolizes sexual arousal. Machinery represents masculine organs. The number three sometimes stands for the penis and testicles. Medusa or a spider may symbolize the aggressive, phallic mother, and a fear of spiders often represents the fear of incest and the abhorrence of the female sexual organs. Sliding, gliding, pulling, and the like may represent masturbation, while ascending stairs, climbing, riding, shooting, dancing, mounting a ladder, rhythmical movements, and some violent motions represent sexual intercourse. Falling out and extraction of teeth in dreams may symbolize castration as a punishment for masturbation.

The regressive nature of dreaming means that it includes memories retained in the unconscious, and some of these memories are death wishes toward people who did not gratify the primary demands of the infant. These death wishes may have been directed toward competing siblings or, as we shall see later, toward one or sometimes both parents. Other such memories are concerned with the early sexual desires of the infant. As the individual grows and matures, normal, adult sexuality develops, but residues of infantile perversions are often included in the dreams of adults.

Along with rejected or repressed wishes, some dreams also carry vehement inner conflicts. What represents a gratification for the unconscious may be perceived as a threat to the preconscious or conscious. In anxiety dreams, for example, the latent dream thoughts undergo little change, and the instinctual demands are too powerful to be warded off. In a nightmare, the threat to consciousness is so great that sleep is disrupted. More often, however, the dreamwork results in *dream distortions* that are the result of a compromise between the unconscious forces that press for a discharge of energy and the opposing force of consciousness, the dream censor, that inhibits, restrains, and counteracts these pressures. The dream censor is the same force that represses the unconscious wishes, keeps them repressed, and resists their expression in free associations necessary for dream interpretation in psychoanalytic therapy.

Somatic Aspects

Somatic stimuli during sleep may elicit additional material for the dream. Sometimes, somatic sensations combine with the mental stimuli and the resulting dream content reflects both sources. Sometimes, the dream is used to deny the physical stimuli or to interpret them away as to make them compatible with sleep instead of disrupting it. "The dream is the guardian of sleep," Freud wrote (1938:287), for the wish to sleep is general, and usually it is combined with other wishes, which are fulfilled in the dream.

Some Typical Dreams

The Dream of Being Naked. The dreamer is naked or scantily clad in the presence of strangers. He is ashamed and embarrassed, and would like to escape or hide, but feels that he cannot move. Usually, the strangers pay no attention to the dreamer's nakedness. This dream reflects a memory of the dreamer's early childhood. Little children are seen naked and are not ashamed or embarrassed. Often, children wish to display themselves, and may experience some degree of excitement when they are naked. The early childhood feelings are reproduced in this dream, and their repetition is a wish fulfillment. "Dreams of nakedness, then, are exhibition dreams." (Freud 1938:294). The embarrassment and shame of the dreamer are caused by repression and the manifest dream content. The feeling of not being able to move is a conflict, for, unconsciously, the dreamer wishes to exhibit himself, but the censor acts against it.

The Dream of the Death of a Beloved Person. The dreamer may be uninvolved or feel profoundly grieved. A dream of the former type disguises a wish which gives no reason for remorse or grief. A dream of the second type reflects a death wish of a beloved person. Such a dream does not necessarily mean that the dreamer wishes that beloved person to die, but that in the past the dreamer had wished him dead. A little child wants his whims to come through despite parental prohibitions and sibling rivalry. Adults may harbor infantile, unconscious, hostile wishes and express them in dreams.

The dreams of the death of beloved persons usually represent a conflict between the repressed wish and the censor. Since the death wish is usually well repressed and the dreamer is unaware of it, the dream censor is unprepared to stop it and thus the wish comes through in a direct and forceful manner. The repressed wish is usually represented in the day's experiences by over and often excessive concern for the beloved person's life. Since in these dreams the repressed wish has avoided the censor, the dream evokes painful emotions.

Dreams of Flying or Falling. These dreams refer to childhood games involving rapid motion, such as swinging, rocking, climbing. Sexual sensations are frequently elicited by these games. Childhood experiences are repeated in these dreams, but their sexual undercurrent evokes anxiety.

Errors of Everyday Life

Everyone makes mistakes, but repetition of error, Freud felt, could hardly be attributed to chance factors, especially when the erroneous behavior is contrary to the conscious wish of the individual. The most common example of this kind of phenomenon are slips of the tongue where we wish to say one thing and say something different. Freud wrote about many such cases: a soldier who said to a friend, "I wish there were a thousand of our men *mortified* (instead of fortified) on that hill, Bill"; a woman who complained that she had an "incurable infernal (internal) disease"; a professor who said in his introductory lecture, "I am not inclined (*geneigt* instead of *geeignet,* that is, fitted) to estimate the merits of my predecessor"; the speaker of a parliament who opened a session with the words, "Gentlemen, I declare a quorum present and herewith declare the session closed" (Freud 1938:31ff).

Some people are aware of the inner conflict and sense it before their tongue slips, while others say that they knew that they had made the slip, but deny that they were aware of it beforehand. Still others not only deny their inner conflict before the slip, but are also unaware of having made the slip.

Slips of the pen often follow the same pattern as slips of the tongue. For example, a writer may make a slip of the pen to let the reader know of something that is on the writer's mind. *Misreading* seems to represent substitution. *Forgetting* of decisions indicates inner opposition or resentment. *Losing or mislaying* objects may be due to an object's having lost its value, being damaged, or representing an unpleasant memory of someone; it may also be an impulse to sacrifice something to hate in order to avert a dreaded loss of something more important. *Forgetting of names* is often related to association of the forgotten name with some unpleasant personality traits of the one who forgot the name.

More generally, the motive for the forgetting of past experiences reflects an unwillingness to recall something that may evoke painful feelings. In the forgetting of resolutions, for example, the conflict that led to the repression of the painful memory becomes tangible. Analysis of such cases points to a strong inner opposition that failed to put an end to the resolution. Some similar conflict is often identified in erroneously carried out actions. The impulse that manifests itself in the disturbances of the actions is frequently a counterimpulse. Often one expresses one's unconscious and repressed wishes through an erroneous action. The acts of taking the wrong train and getting out at the wrong station illustrate the point.

We see then that the same condensations and the same compromise formations *(contaminations)* that are found in dreams are also found in the errors of daily life. The situation is quite similar, since unconscious thoughts find expression in a disguised fashion. The incongruity, irrationality, and absurdity of the manifest dream correspond to the irrationality of unconscious wishes that express themselves in the errors of everyday life.

Symptom Formation

Freud defined neurotic symptom formation as a compromise between the unconscious impulses and the reality-oriented censorship. A symptom represents a fulfillment of an unconscious wish, distorted by the same censoring forces that oppose the fulfillment of the wish. Freud distinguished a clear difference between symptom formation and dream formation, however:

> The preconscious purpose in dream-formation is merely to preserve sleep and to allow nothing that would disturb it to penetrate consciousness.

> It does not insist upon confronting the unconscious wish-impulse with a sharp prohibiting "No, on the contrary." It can be more tolerant because a sleeping person is in a less dangerous position; the condition of sleep is enough in itself to prevent the wish from being realized in actuality. (Freud 1915–1917:315)

Like errors and dreams, neurotic symptoms have a hidden meaning and they are closely connected with traumatic events in the patient's life.

A traumatic experience may be defined as an experience that, within a very short space of time, subjects the mind to such an intense degree of stimulation that assimilation or elaboration of it cannot be effected by normal means, and lasting disturbances in the distribution of the available mental energy result. A great effort is needed to prevent the traumatic and disturbing process from entering into consciousness. In the end, the trauma is pushed into the unconscious or repressed, but there, as an unconscious conflict, it has the power to develop symptoms.

Freud discovered that the main area of repression in symptom formation is related to sexual experiences and wishes. He found that experienced *or imagined* sexual traumas are the chief cause of symptoms. Of course, the precise cause of a trauma cannot always be ascertained, but the earlier a trauma takes place, the more serious is the resultant damage to the personality. Resolution of hidden, traumatic conflicts is one of the main objectives of psychoanalytic treatment, and each one of Freud's treatment successes added support to his theory of the unconscious.

Telepathy

Freud maintained that whatever exists has a cause and an effect and must not be dismissed. Even wit and jokes did not escape Freud's keen eye; for him, a joke gave vent to wishes that could not otherwise be expressed. Another area that was usually bypassed by research workers was telepathy, and Freud took an empirical approach to the study of it. He wrote: "First, we have to establish whether these processes really occur, and then, but only then, where there is no doubt that these processes really occur, we can set about their explanation" (Freud 1932:40). For Freud, the fact that charlatans busy themselves with occult phenomena did not prove that these phenomena do not exist.

Freud cited several instances of telepathic dreams, intuitive knowledge, and transference of thought, and he explained them in the following manner. He felt that certain individuals have a better access to their unconscious than others do, and that these people seem to be somehow aware of their unconscious feelings about others and of the feelings of other people about them. These people may, for example, display unexplainable foresight that is somehow related to the anticipation of risks and dangers. At any rate, Freud believed that these deep layers of the unconscious required a cautious approach and objective investigation.

BIBLIOGRAPHY

Arlow, J. and C. Brenner. 1964. *Psychological Concepts and the Structural Theory*. New York: International Universities Press.

Freud, S. Standard Edition. James Strachey, ed. 24 vols. London: Hogarth, 1962; New York: Macmillan, 1962.

—— 1887–1902. *The Origins of Psychoanalysis: Letters to Wilhelm Fliess, Drafts and Notes*. New York: Basic Books, 1954.

—— 1895. *Project for Scientific Psychology*. Standard Edition, 1:283–398.

—— 1900. *The Interpretation of Dreams*. Standard Edition, vols. 4 and 5.

—— 1905. *Jokes and Their Relation to the Unconscious*. Standard Edition, 8:9–238.

—— 1912. *A Note on the Unconscious in Psychoanalysis*. Standard Edition, 12:255–262.

—— 1915. *The Unconscious*. Standard Edition, 14:159–204.

—— 1916. *Introductory Lectures on Psychoanalysis*. Standard Edition, 15:15–242.

—— 1920. *Beyond the Pleasure Principle*. Standard Edition, 18:7–64.

—— 1923. *The Ego and the Id*. Standard Edition, 19:3–63.

—— 1933. *New Introductory Lectures on Psychoanalysis*. Standard Edition, 22:3–84.

—— 1938. *An Outline of Psychoanalysis*. New York: Norton, 1949.

Gill, M. M. 1963. Topography and systems in psychoanalysis theory. *Psychological Issues*, 3:2.

Holt, R. R. 1962. A critical examination of Freud's concept of bound versus free cathexis. *Journal of the American Psychoanalytic Association*, 10:475–525.

—— 1976. Freud's theory of primary process—present status. In T. Shapiro, ed., *Psychoanalysis and Contemporary Science*, 5:61–99. New York: International Universities Press.

Pavlov, I. P. 1928. *Lectures on Conditioned Reflexes*. New York: Liveright.

Sulloway, F. J. 1983. *Freud: Biologist of the Mind*. New York: Basic Books.

Wolman, B. B., ed. 1979. *Handbook of Dreams: Research, Theory, and Applications*. New York. Van Nostrand-Reinhold.

Zilboorg, G. and C. W. Henry. 1941. *A History of Medical Psychology*. New York: Norton.

[FIVE]
The Driving Forces

THE previous chapter described the unconscious determinants of normal and abnormal personality development. The present chapter discusses motivation and the dynamic determinants of normal and abnormal behavior.

Principle of Constancy

Every organism, Freud wrote, tends to retain an equilibrium unless exposed to stimulation. This principle of equilibrium, called by Cannon *homeostasis,* was called by Freud the *principle of constancy.* According to this principle, every organism exposed to a stimulus that disrupts the equilibrium reacts in a manner aimed at the restoration of the initial state. The entire behavioral process can be presented as a continuum of equilibrium: stimulus that causes tension—action (discharge or energy)—restoration of balance that brings relief.

Under the influence of Fechner, Freud believed that tension corresponds to pain and relief is identical with pleasure. Avoidance of pain and pursuit of pleasure became the framework of Freud's "economic" theory that viewed human behavior as a tension-relief, or pain-pleasure continuum (Freud 1920).

However, this simple behavioral model could not explain the complexity of human behavior. According to Freud, there are distinct driving and inhibiting forces that respectively facilitate and prevent the discharge of mental energy called libido.

Instinctual Forces

Originally Freud believed that all living organisms act in accordance with two purposes, self-preservation and the preservation of the species.

Accordingly, Freud distinguished between self-preservation or ego instincts and sexual instincts. These two instinctual forces are often in conflict with each other. The sexual instincts are more flexible than the self-preservation instincts; they can be held in suspense (aim inhibited), sublimated, diverted into new channels, distorted, and perverted; their gratification can be denied or substituted for and their objects can be easily changed.

Freud did not postulate a single drive that leads to fertilization and preservation of the species. He believed that there are a number of relatively independent instincts that stem from various somatic sources. All of them strive toward gratification in their respective somatic zone, or to *organ pleasure*. In the process of ontogenetic development some of them merge with the sexual instinct proper that originates in the genital organs; some of these instincts will be eventually repressed, some partially incorporated in the final organization of the adult sexual functions.

Originally Freud assumed that both normal and abnormal sexuality stem from the same source, and therefore he had to conclude that sexual deviations are some sort of retardation or thwarting of the sexual development. This was a far-reaching hypothesis. Comparative studies in biology and embryology led him to believe that Haeckel's biogenetic theory can be readily applied to the development of sex. "Ontogenesis is a recapitulation of phylogenesis," Haeckel said, and Freud, in his essays about sex, accepted this theory and assumed that the child's sexual development is a recapitulation of the main phases in the evolution of sex in organic nature (Freud 1905).

Infantile Sexuality

The study of sexual perversions in adults and the biogenetic principle led to the study of infantile sexuality. "The child is psychologically father of the man," Freud said. Infantile sexuality contains all the potentialities for the future development which may lead in any direction—either into a normal or into an abnormal sexuality. Normal development takes place when the source, the object, and the aim of sex are combined in a consistent effort of unification of genital organs of two persons of the opposite sex. The genitals are the normal sexual source; the object is an adult person of the opposite sex; the aim is heterosexual intercourse. The normal individual passes through developmental stages and if, for some reason, he retains the characteristics of one of them (remains "fixated"), he is considered abnormal. What is normal in infancy is abnormal in adulthood. A sexually abnormal individual is a sexually retarded individual. The infant is "a polymorphous pervert" (i.e., interested only in obtaining pleasure) that may or may not eventually become a well-adjusted adult.

Ego and Libido

Later on, Freud modified his theory of instincts. Despite changes and revisions, Newton's physics and Darwin's evolution have always served as Freud's frame of reference. Inertia was his basic law; that is, bodies were inert unless exposed to forces that caused movement. The forces acting in organic nature were called *Triebe* or drives (sometimes translated as instinctual drives or instincts). These forces are inherited and are geared toward two main goals: preservation of individual life and preservation of the species. Originally, Freud called the first *ego drives* and the latter *libido drives*. In 1914, however, Freud introduced the idea of narcissism and modified the theory of instinctual drives.

The ego drives toward self-preservation help the individual adjust to life. The libido, or sexual drives, on the other hand, are more impulsive, and it takes years before they become at least somewhat subordinated to reality considerations. The self-preservation drives are not too flexible. One cannot indefinitely postpone, for example, the gratification of hunger or thirst, and there is not much latitude for change in the ways one can satisfy his thirst or hunger. Sexual instincts, however, can be altered with regard to the bodily zone, aim, and person with whom one seeks gratification. The self-preservation drives are relatively clear-cut, while the sexual drives are full of conflicts, substitutions and deviations, and perversions.

Primarily, the modifications in sexual drives are related to aims and objects. Transformations, fusions, and substitutions of instinctual gratifications are quite frequent and often apparent. Freud's analytic experience gave him evidence that instinctual impulses from one source can join onto instinctual impulses from another and share their further vicissitudes, and that in general the satisfaction of one instinct can be substituted for the satisfaction of another. Freud also found that the relations of an instinct to its aim and to its object are susceptible to alterations. Both can be exchanged for others, but the relation to the object is the more easily loosened of the two (Freud 1933:133). In other words, the energy at the disposal of an instinct can be *cathected* (invested) into a certain person, withdrawn, and reinvested in another.

In some cases the instinctual drive comes to a stop at a certain point and renounces its full gratification. This takes place when the instinctual drive becomes too powerfully cathected in a certain object. For instance, affection represents such a case of permanent object-cathexis; that is, one that expresses itself in a constant caring for the beloved person without sexual gratification. Freud called this process of suspension of gratification *aim-inhibition*. A genuine parental love is aim-inhibited in that loving parents do not expect sexual gratification from their children.

Still another modification takes place when libido is redirected from the search for sexual gratification into a socially useful channel. Freud called this modification *sublimation*. Freud believed that creative art was an example of the process of sublimation. The energies put originally at the disposal of sex are redirected in another channel of creative work in art, music, or literature (Gedo and Pollock 1975).

Sexual Life

According to the pleasure principle, people are bent upon procuring pleasure and avoiding pain. Freud wrote: "We may venture to say that pleasure is in some way connected with lessening, lowering, or extinguishing the amount of stimulation present in the mental apparatus; and that pain involves a heightening of the latter. Consideration of the most intense pleasure of which man is capable, the pleasure in the performance of the sexual act, leaves little doubt upon this point" (1916:311).

Ten years earlier Freud wrote:

> The popular view distinguishes between hunger and love, seeing them as representatives of the instincts that aim at self-preservation and reproduction of the species respectively. In associating ourselves with this very evident distinction we postulate in psychoanalysis a similar one between the self-preservative or ego-instincts on the one hand and the sexual instincts on the other; the force by which the sexual instinct is represented in the mind we call "libido"—sexual longing—and regard it as analogous to the force of hunger, or the will to power, and other such trends among the ego-tendencies. . . .
>
> We have defined the concept of libido as a quantitatively variable force which could serve as a measure of processes and transformations occurring in the field of sexual excitation. We distinguish this libido in respect of its special origin from the energy which must be supposed to underlie mental processes in general, and we thus also attribute a qualitative character of it. (Freud 1905:217)

For Freud, there was no single streamlined drive that leads to fertilization and preservation of the species. "Sexual life," he states, "comprises the function of obtaining pleasure from zones of the body" (Freud 1938:26). There have always been several instinctual needs related to various somatic sources. In the process of ontogenetic development, some of them have merged with the proper sexual instinct and, subsequently, brought into the service of the function of reproduction. However, in some cases the two functions fail to coincide completely.

It is necessary to distinguish between source, object, and aim in the sexual drive. The *source* is a stimulation arising in some part or zone of the organism. The various parts of the body that react to sexual stimuli, such as the genitals, the mouth, and the anus, are called *erotogenic zones*. The usual *object* of sexual desires is an adult person of the opposite sex; but a person of same sex is the object for homosexuals.

Psychosexuality

In Freud's terms sexuality includes far more than it does as generally used. This extension is justified genetically. In psychoanalysis, all expressions of tender feeling are seen as belonging to "sexual life," since these feelings presumably spring from the source of primitive sexual feelings, even when they have become inhibited in regard to their original sexual aim or have exchanged this aim for another that is no longer sexual. Therefore, the term *psychosexuality* is preferred in psychoanalysis, thus laying stress on the point that the mental factor should not be overlooked or underestimated.

According to Freud, the newborn child already carries the germs of sexual feelings which continue to develop for some time and then succumb to a progressive suppression, which is in turn broken through by the proper advances of sexual development and which can be checked by individual idiosyncrasies (Freud 1905c). In addition, *infantile amnesia* causes most people to reject this contention and view their childhood years as if they belonged to a prehistoric time.

In actuality, infantile amnesia is a process of repression. Infantile sexual drives are incompatible with adult life and, as such, they must remain deeply buried in the unconscious. These hidden sexual desires often appear in dreams, but in a disguised manner. The first and most pleasurable activity of an infant is sucking from the mother's breast, and the nipple of a milk bottle can be viewed as the prototype of all future pleasures. "He who sees a satiated child sink back from his mother's breast and fall asleep with reddened cheeks and blissful smile, will have to admit that this picture remains as typical of the expression of sexual gratification in later life" (Freud 1905c). Children who retain the action of thumb sucking for a prolonged period of time may show a tendency as adults to intense or even perverse kissing, and experience the desire for excessive drinking and smoking.

Abnormal Sexuality

Freud's concept of sexuality included perversions and infantile sexuality that do not lead to the usual aim of unification of the genital organs in the act of intercourse. A perverse sexual relation may end with orgasm and ejaculation analogous to normal sexual intercourse. The forepleasure activities of normal individuals include several elements that, if performed exclusively, and in place of normal intercourse, should be considered perverse. Although kissing is an indispensable part of the sexual foreplay of normal individuals, it is primarily a contact of oral zones. The mouth is an erotogenic zone and is therefore capable of producing sexual excitation. Some individuals obtain orgasm by kissing, "deep kissing," or by oral-genital contact. According to Freud, if kissing prevents normal sexual intercourse, it is perverse. Thus normal and perverse sexuality have the same roots.

The infant is a "polymorphous pervert" who must grow and develop to become normal. He must modify some of his early sexual preferences, some must be discontinued, and still others must be incorporated into sexual foreplay and subordinated to the adult sexual aim of unification of the sexual organs. Onlooking is a widely practiced element of sexual foreplay which becomes perverted voyerism only if its practice excludes the union of genitals.

Sexual deviations are not a "degeneration," but are a sort of thwarting or regression of sexual development. Infantile sexuality contains all the potentialities for future development in any possible direction: toward normal sexuality, or becoming fixated on one of the early stages or functions.

> If we are led to suppose that neurotics conserve the infantile state of their sexuality or return to it, our interest must then turn to the sexual life of the child, and we will then follow the play of influences which control the processes of development of the infantile sexuality up to its termination in a perversion, a neurosis, or a normal sexual life. (Freud 1905)

Narcissism

"There are various points in favor of the hypothesis of a primordial differentiation between sexual instincts and other instincts, ego-instincts. . . . This differentiation of concepts corresponds to the distinction between hunger and love" (Freud 1914:77). After some years of clinical experience Freud found cases in which libido is directed toward the self and not toward external objects only. In fact, he concluded, love-for-oneself must precede object love. "The sexual instincts are at the outset supported upon the ego-instincts [for] the first autoerotic sexual gratifications are experienced in connection with vital functions in the service of self-preservation" (p. 84). It takes time for infants to learn to divert part of the love primarily cathected (invested) in themselves and to cathect it in their mothers. Freud called this self-love *narcissism* after the legendary Greek youth Narcissus, who fell in love with himself (Gedo and Pollock 1975).

Narcissism starts in prenatal life and never really disappears. At the earliest stage of life it is the libido's only channel of cathexis, as all mental energies are invested in oneself. This initial stage Freud called *primary narcissism*. Later, when object love is thwarted, the libido may turn back to one's own person and *secondary*, morbid narcissism may develop.

Eros

The discovery of the phenomenon of narcissism destroyed for Freud the barriers separating the libido from the ego instincts. From this point on, he considered the ego instincts to be special cases of libido cathexis, namely as an investment of libido in one's own person. Freud was left with but one instinctual force, the force of love, the libido, that can be cathected (invested) in oneself in narcissistic love or in others, in object love.

The conflict between sex and self-preservation instinct drives was reinterpreted by Freud as narcissism versus object love. To be well adjusted, an individual must have a balanced distribution of libido cathexes between oneself and others. A certain amount of narcissism is necessary purely for self-protection, but if one also is to take care of others, one must be capable of object cathexis as well. Many maladjusted individuals suffer from an imbalance in cathexes. Some of these develop secondary and morbid narcissism after they have been seriously thwarted in the development of object cathexis, while others are unable to take care of themselves due to insufficient narcissism or abundant object cathexis.

Freud united under the name of *Eros* all the forces that serve pleasure and enhance the vital functions of the individual. Eros encompassed all sexual and egoistic drives, and libido became the name for all energies, whether self or object divested, that are at the disposal of erotic force.

Sadism and Masochism

Freud's theory of Eros and libido could not explain the senseless destruction that had taken place down through the ages. It could explain attacks for gain and profit, but it could not explain senseless brutality and murder for their own sake.

The matter became even more complicated for Freud when he attempted to explain sexual brutality. Some men seem to be unable to enjoy the sex act unless they inflict pain on their female partner. This behavior, called sadism, cannot be interpreted by the theory of Eros because it is not an act of love and affection. Love does not make people inflict pain on their love objects. The empirical fact of sadism forced Freud to reconsider his theory of instincts.

Masochism was perhaps even more difficult to understand. How could one combine love with the wish to be hurt? How could pain be a prerequisite of the greatest sensual pleasure? After all, the desire to be hurt could easily lead to self-destruction. But then, suicidal impulses, so frequently met in mental patients, could not be interpreted in terms of the libido theory either. Libido could serve self-preservation or the preservation of love objects, but it could not be a tool for the destruction of oneself or of others.

Freud also observed that in many instances instinctual processes do not follow the pleasure principle. They appear to be *beyond the pleasure principle*. Patients with traumatic neurosis display a tendency to reexperience in waking life and/or in dreams their earlier emotional experiences, despite the fact that they perceive repeated experiences as anxiety provoking. As a result of these observations, Freud (1920) elaborated upon the *conservative nature* of instincts and redefined instincts as a tendency innate in living organic matter "impelling it towards the reinstatement of an earlier condition." From this premise, Freud deduced the existence of destructive-aggressive energies which are part of a primal "death instinct" called Thanatos. Thanatos can be conceived of as being composed of impulses which manifest themselves in terms of *regressions to earlier modes of personality integration, suicide and self-destructive behavior patterns*, and *aggressive behavior towards others*.

Nirvana Principle

These phenomena of repetition compulsion represent the *nirvana principle*. Nirvana is the tendency of mental life to reduce to nothing or, at least, to maintain at "as low a level as possible the quantities of excitation flowing into it." The nirvana principle is somewhat analogous to the pleasure principle, the regulator of eros and the libido. Freud believed that the nirvana principle is primal. The inorganic (not-life) state existed prior to the organic (life) state, therefore, Freud postulated that nirvana precedes the pleasure principle and wrote as follows:

> We have taken the view that the principle which governs all mental processes is a special case of Fechner's tendency towards stability; and have accordingly attributed to the mental apparatus the purpose of reducing to nothing, or at least of keeping as low as possible, the sums of excitation which flow in upon it. Barbara Low . . . has suggested the name of 'nirvana principle' for this supposed tendency, and we have accepted the term. But we have unhesitatingly identified the pleasure-unpleasure principle with this nirvana principle . . . the nirvana principle (and the pleasure principle which is supposedly identical with it) would be entirely in the service of the death instincts, whose aim is to conduct the restlessness of life into the stability of the inorganic state, and it would have the function of giving warnings against the demands of the life instincts—the libido—which try to disturb the intended course of life. . . . We must perceive that the nirvana principle, belonging as it does to the death instinct, has undergone a modification in living organisms through which it has become the pleasure principle; and we shall henceforward avoid regarding the two principles as one. It is not difficult, if we care to follow up this line of thought to guess what power was the source of the modification. It can only be the life instinct, the libido, which has thus, alongside the death instinct, seized upon a share in the regulation of the processes of life. In this way we obtain a small but interesting set of connections. The *nirvana* principle expresses the trend of the death instinct; the pleasure principle represents the demands of the libido. (Freud 1920:159–160)

Aggression

As early as in 1897, Freud first wrote about hostile impulses against parents (a wish that they might die) and he described them as an integral part of neuroses. He found that they came to light consciously in the form of obsessional ideas. "In paranoia the worst delusions of persecution . . .

correspond to these impulses. They are repressed at periods in which pity for one's parents is active—at times of their illness or death. One of the manifestations of grief is then to reproach oneself for their death . . . or to punish oneself in a hysterical way by putting oneself in their position with the idea of retribution" (Freud 1887–1902).

In 1908, Alfred Adler suggested a separate aggressive drive (Loewenthal 1983; Nunberg and Federn 1962:408). Although Freud never denied the existence of aggressive behavior, he linked aggression to libido, and could not accept Adler's notion of self-assertion as the underlying principle of aggression. His own concept of aggression was anchored to the general principle of equilibrium. Nature tends to retain its *status quo;* if the original equilibrium is disturbed, the tendency is to go back and to restore the initial state.

> Let us make a sharper distinction that we have hitherto made between function and tendency. The pleasure principle, then, is a tendency operating in the service of a function whose business it is to free the mental apparatus entirely from excitation or to keep the amount of excitation in it constant or to keep it as low as possible. We cannot yet decide with certainty in favor of any of these ways of putting it; but it is clear that the function thus described would be concerned with the most universal endeavor of all living substance—namely, to return to the quiescence of the inorganic world. (Freud 1920:62)

In this way Freud justified the need of a new theory of instincts.

> With the hypothesis of narcissistic libido . . . the sexual instinct was transformed for us into Eros, which seeks to force together and hold together the portions of living substance. . . . Our speculations have suggested that Eros operates from the beginning of life and appears as a "life instinct" in opposition to the "death instinct" which was brought into being by the coming to life of the inorganic substance. These speculations seek to solve the riddle of life by supposing that these two instincts were struggling with each other from the very first. (*Ibid.*, note 55)

Thanatos

Based on the fact that organic matter developed from inorganic matter, Freud believed that with the start of life an instinct was born that aimed at the return of the inorganic state and destruction of life. The aim of this death instinct, called *Thanatos,* was the reestablishment of inanimate nature. Life ends in death, and death leads to new life. Thus for Freud, Eros and Thanatos are interwoven, and construction and destruction are inseparable. Even the process of life itself cannot be entirely free from the death instinct. The erotic instincts try "to collect living substance together into even larger unities, (and the death instincts act against this) and try to bring living matter back into inorganic condition. The cooperation and opposition of those two forces produce the phenomena of life to which death puts an end" (Freud 1933:146–147).

> After long doubts and vacillations we have decided to assume the existence of only two basic instincts, *Eros* and the destructive instinct. . . . The aim of the first of these basic instincts is to establish ever greater unities and to preserve them, thus—in short, to bind together; the aim of the second, on the contrary, is to undo connections and so to destroy things. We may suppose that the final aim of the destructive instinct is to reduce living things to an inorganic state. For this reason we also call it the death instinct. (Freud 1938:20)

Repetition—Compulsion

In *Beyond the Pleasure Principle* (1920), Freud introduced a series of new concepts deviating substantially from his early theory of motivation. The constancy principle, or the restoration of the initial state, became more important than the search for pleasure. Freud now believed that people reexperience past tensions in dreams not because the dreams bring relief and pleasure, but because the dreamers are compelled to repeat past experiences. *Repetition compulsion* is the inescapable product of the principle of constancy. Past tensions, pleasurable or painful, tend to undergo repetition until they are resolved and an equilibrium is established. By means of the repetition compulsion, Freud explained the tendency to reproduce painful and traumatic experiences; that is, that the instinctual forces produce continuous repetitions of the disturbing experiences until a balance is restored.

Freud explained that one's natural aggressiveness against the self may be directed against the outer world. He believed that people have to destroy things and other people in order not to destroy themselves. In order to protect oneself from the tendency toward self-destruction, one must find external channels for aggressiveness.

At this point in Freud's thinking, all instincts were directed toward the reinstatement of an earlier state of things. When the given state is upset, an instinctual action starts, aiming at the restoration of that state. In such an action, Eros and Thanatos may combine their resources, but no less often they fight each other. Eating, for example, is a process of destruction with the purpose of incorporation; sexual intercourse is an act of aggression that aims at the "most intimate union." Sexual impulses are rarely purely erotic, quite often combining erotic and destructive demands. When the destructive instincts become the stronger part in the fusion of the two kinds of instincts, sadism or masochism results.

When an individual's behavior, under the control of Thanatos, is directed toward the outer world, he becomes hateful and aggressive, spreading destruction and death. When these forces are directed against oneself, self-defeat and self-destruction (suicide) may put an end to life.

Theory of Instincts

As mentioned above, all these theoretical changes were related to the concept of homeostasis or equilibrium, called by Freud the constancy principle. This idea, originated in Newton's law of inertia, was applied by Fechner to neurology under the name "tendency toward stability" and elevated by Freud to the role of a guiding principle in psychology. Freud discovered the tendency to repetitive behavior, irrespective of pleasure or displeasure. This repetition-compulsion has pointed to the "*conservative* nature of living substance" (1920).

Thus, Freud hypothesized that

> all organic instincts are conservative, are acquired historically, and tend towards the restoration of an earlier state of things. It follows that the phenomena of organic development must be attributed to external disturbing and diverting influences. . . . It would be in contradiction to the conservative nature of the instincts if the goal of life were a state of things which had never yet been attained. If we are to take it as a truth that knows no exception that everything dies for *internal* reasons—becomes inorganic once again—then we shall be compelled to say, that "the goal" of life "is death" and, looking backwards, that what was inanimate existed before what is living. (1920:49–50)

With these modifications the ground was laid for a new theory of personality structure to be described in detail in chapter 7. This new theory, which introduced the tripartite model of id, ego, and superego, incorporated the two earlier ideas of mental strata and dynamic forces. The id was postulated as a reservoir of instinctual impulses which continually seek immediate gratification in accordance with the pleasure principle. The id's cathectic mobility could be in the processes of displacement and condensation. The id's primary processes are full of unresolved contradictions and archaic symbolisms. The id has no perception of external events and is, therefore, incapable of learning. The unconscious does not change and the repressed infantile memories last forever and manifest themselves in dreams, parapraxes and in symptom formation.

The two primal instincts, Eros (the life instinct) and Thanatos (the death instinct), have their seat in the id. Libido, the energy of Eros, is regulated by the pleasure principle and preservation of life.

Thanatos and its destructive and regressive impulses, manifested in repetition compulsion, are regulated by the nirvana principle. The aim of the nirvana principle is to eliminate excitations and somatic-psychic tension. Although Eros and Thanatos represent opposing instinctual forces, they may merge together on many occasions, eating being one of them. The fusion of the destructive instinct and the libido in the psychosexual development of children takes place in the "oral-sadistic" and the "anal-sadistic" stages to be described in chapter 6.

BIBLIOGRAPHY

Freud, S. Standard Edition. James Strachey, ed. 24 vols. London: Hogarth, 1962; New York: Macmillan, 1962.

—— 1887–1902. *The Origins of Psychoanalysis: Letters to Wilhelm Fliess, Drafts, and Notes*. New York: Basic Books, 1954.

—— 1900. *The Interpretation of Dreams*. Standard Edition, vols. 4 and 5.

—— 1905. *Three Essays on the Theory of Sexuality*. Standard Edition, 7:125–243.

—— 1914. *On Narcissism: An Introduction*. Standard Edition, 14:73–102.

—— 1916. *Introductory Lectures on Psychoanalysis*. Standard Edition, 15:15–42.

—— 1920. *Beyond the Pleasure Principle*. Standard Edition, 18:7–64.

—— 1933. *New Introductory Lectures on Psychoanalysis*. Standard Edition, 22:3–184.

—— 1938. *Outline of Psychoanalysis*. New York: Norton, 1949.

Gedo, J. E. and G. H. Pollock. 1975. Freud, the fusion of science and humanities: The intellectual history of psychoanalysis. *Psychological Issues* 9, (Monographs 34–35).

Loewenthal, H. S. 1983. Psychoanalysis in central Europe. In B. B. Wolman, ed., *International Encyclopedia of Psychiatry, Psychology, Psychoanalysis, and Neurology: First Progress volume*. New York: Aesculapius Publishers.

Nunberg, H. and E. Federn. 1962. *Minutes of the Vienna Psychoanalytic Society*. New York: International Universities Press.

[SIX]
Developmental Stages

BEING born is a traumatic experience which disrupts the well-balanced life of the organism in the uterus. Birth, as the first trauma, is the prototype of all anxiety feelings in later life. The organism is flooded by stimuli, tensions come to a peak, and the helpless organism is exposed to the shock of being born.

The Neonate

The newborn's mental apparatus is exposed to stimuli far beyond his capacity to handle. The natural tendency is therefore to restore the mental balance through a withdrawal from reality—falling asleep. Once an unpleasant tension, such as hunger, is removed, the infant falls asleep with a feeling of profound happiness and bliss. Only when stimulated by hunger or cold or some other discomfort is he awake. The neonate is completely narcissistic. In this stage of primary narcissism objects in the outer world are barely noticed and the infant is unable to distinguish between wish and reality, self and the outer world. Gratification of his needs comes immediately, and he may feel as "omnipotent" when a wish is gratified (after receiving food or after eliminations) as he felt miserable when the tension arose.

The infant's earliest attitude to the outer world can be characterized as an objectless longing, or some sort of unconscious craving for unification with the outer world. Freud called this foggy longing the *oceanic feeling*. It is as if the infant were longing to go back to the uterus, or, even farther, to nonexistence. Eros and Thanatos are united in the striving toward the all-encompassing and soothing passivity which ultimately means death.

The first pleasure-giving objects which the infant meets are the nipples of the mother's breast or the bottle. The first somatic zone to experience pleasurable sensation is his mouth. Thus the oral stage of instinctual development begins immediately after birth (Freud 1905).

The Oral Phase

It is assumed that the child's sexuality is polymorphously perverse and leads through developmental stages to the adult and normal sexuality. One of the main supports for this assumption is the biogenetic principle. Additional reasons for this assumption are derived from clinical studies indicating that normal and perverse sexuality develop from the same source, infantile sexuality. Both normal and perverse sexuality aim at the same goal—orgasm; both stem from the unorganized infantile sexuality, where various wishes and desires exist independently. In adult sexuality one component becomes dominant, and the entire sexual activity becomes concentrated in one area, the genital zone, other areas being excluded or relegated to secondary roles. The genital zone is dominant in normal adults; in individuals with perversions, some other zone is dominant. In childhood, any area or all areas may strive for their own pleasure. Some adults whose sexuality remains infantile remain polymorphously perverse.

The development of each individual depends upon innate instinctual forces, biologically determined developmental stages, and environmental influences. Freud never underestimated the environmental factors. How quickly, how successfully, how completely the individual passes the developmental stages, and how much of them is carried over into his adult life, depends mainly upon the interaction of the child with his environment.

The earliest sexual life of an infant consists of a series of independent activities of single-component impulses each seeking pleasure in a bodily organ, i.e., organ pleasure. These various organs are *erotogenic zones*. They serve as centers for sexual gratification in the pregenital stages of libido development, prior to the subordination of all the sexual component instincts under the primacy of the genital zone.

The infant's first sexual excitations are connected with the feeding process. As the infant falls asleep at the breast, completely satisfied, "it bears a look of perfect content which will come back again later in life after the experience of the sexual orgasm." The infant may continue to suck even if he does not take any food; he is sucking for the pleasure of sucking.

Sucking for nourishment is the "prototype of every later sexual satisfaction." The desire to suck includes the desire for the mother's breast, which is therefore the first *object* of sexual desire. Love is attached to hunger, and sucking brings gratification both of hunger and of love. At the beginning the child does not distinguish between the mother's breast and his own body. As he turns to sucking for pleasure, the breast is given up as a love object and is replaced by a part of his own body; he sucks his own thumb or tongue, and his own body is his love object.

Whether the child was breast fed or bottle fed, weaned early or late, he always longs for the mother's breast and for the mother as a whole. She stays in his memory as his first love object, and his feelings for her are the prototype of all love relations. This occurs with both boys and girls.

At the oral phase of libido organization, object love is ambivalent and contains both Eros and Thanatos. The loved object is assimilated by eating or swallowing it, and it is annihilated at the same time. The infant wishes to swallow what he loves and his love leads to swallowing, i.e., to destruction of love objects. It is a cannibalistic tendency; a cannibal, said Freud, has "a devouring affection" for his enemies and devours people of whom he is fond.

Once the milk is swallowed, the infant does not care for it any more. Once tension is removed, he falls asleep and discontinues the contact with the outer world. This is a primitive kind of love—a love for a while, a love that terminates itself immediately after gratification.

Adults who did not outgrow the oral stage may retain these destructive elements in their love in adult life. They love their love objects inasmuch as they can exploit them, and they love only as long as the exploitation goes on.

Identification

One of the infant's earliest emotional ties to the outer world is *primary identification*. This is the wish to be like the other person, and it antecedes a true *object relationship*, which is the wish to possess the other person. Identification is not necessarily love: "It can turn into an expression of tenderness as easily as into a wish for someone's removal."

Identification at this stage is accomplished by *introjection*. The infant takes things into his mouth, thus incorporating whatever he loves. Once he incorporates the object, he may believe himself to be like the beloved object. To love something at the oral stage is identical with the wish to incorporate it or to introject it; oral intorjection is the means for primary identification.

Primary identification, which is a general phenomenon in earliest childhood, must be distinguished from what Freud called *secondary iden-*

tification, which takes place in pathological cases at a later developmental stage. Sometimes a loss of love object causes desexualization of it and giving up of object cathexis; the mourner "introjects" his lost love object, and his object relationship gives way to a regression to identification with the lost love object.

Karl Abraham (1911, 1924) an associate of Freud, suggested dividing the oral phase into *oral-passive* and *oral-aggressive* (see table 6.1). The oral-passive phase extends over several months of the first year of life; toward the second (often well in the second) year, the oral-aggressive phase begins. The oral-passive or oral-dependent stage is characterized by pleasure being derived from sucking. The infant may not be able to distinguish clearly between himself and the external world, and he perceives in the sucking a self-gratifying and a taken-for-granted experience. The oral-aggressive phase usually coincides with teething. The infant becomes aware that the mother's breast is not a part of himself, that it is not available as he wishes it to be. He can no longer take the breast for granted. When frustrated, he forces his way through, he grabs and bites, he tries to receive his oral gratification by acts of aggression.

Eros and Thanatos are combined in the entire oral stage in its canni-

Table 6.1 Abraham's Timetable

Stages of Libidinal Organization	*Stages in Development of Object Love*	*Dominant Point of Fixation*
Early oral (sucking)	Autoeroticism (no object, preambivalent)	Certain types of schizophrenia (stupor)
Late oral-sadistic (cannibalistic)	Narcissism: total incorporation of the object	Manic-depressive disorders (addiction, morbid impulses)
Early anal-sadistic	Partial love with incorporation	Paranoia, certain pregenital conversion neuroses
Late anal-sadistic	Partial love	Compulsion neurosis, other pregenital conversion neuroses
Early genital (phallic)	Object love, limited by the predominant castration complex	Hysteria
Final genital	Love (postambivalent)	Normality

The table follows Fenichel's modifications (Fenichel 1945:101)

balistic tendencies of swallowing. However, during the first part of the oral stage the infant is basically passive, and takes for granted the supply of milk. At the oral-aggressive stage, aggressiveness is utilized as a weapon in procuring gratification. The infant swallows as before, but in addition he may spit and bite, and these are definitely aggressive patterns of behavior.

The Young Child

The Anal Phase

In the second and often third year of life, the child derives considerable pleasure from excretion, and he learns to increase the pleasure by retaining feces and stimulating the mucous membranes of the anus. This second stage of libido development was called by Freud *anal-sadistic.*

> Infants experience pleasure in the evacuation of urine and the content of bowels, and they very soon endeavor to contrive these actions so that the accompanying excitation of the membranes in these erotogenic zones may secure them the maximum possible gratification. . . . The outer world steps in as a hindrance at this point, as a hostile force opposed to the child's desire for pleasure. . . . He is not to pass his excretions whenever he likes but at times appointed by other people. . . . In this way he is first required to exchange pleasure for value in the eyes of others. (Freud 1916)

Ambivalence

The child values his feces as "part of his own body and is unwilling to part with them." He feels that the feces are his property and no one may exercise control over them. Therefore he may offer resistance to the social pressures. He may act aggressively by elimination; libido and hate are combined in anal eroticism in the pleasure of defecating and the sadistic "getting rid" of feces. When the child resists bowel training and holds back feces, he expresses in another way his opposition to adults.

The anal stage is ridden with another ambivalence besides the expulsion retention one. At this stage, masculinity and femininity are distinguished by activity and passivity respectively. Active expulsion of feces is masculine. Masculine impulses are *scotophilia* (gazing), onlooking, curiosity, desire to manipulate and to master, which easily develops into cruelty and sadism. Feminine impulses consist of a passive desire connected with the anal and hollow erotogenic zone. The rectum can be easily stimulated by accepting a foreign body that enters it. The anal ambivalence of masculine-active expulsion and feminine-passive reception of a foreign body may lead to bisexual tendencies in later life (Freud 1908).

At this period the child's sexual interest is primarily focused on the problem of birth, usually caused by the birth of a sibling. Many children believe that babies are born by bowel and are eliminated as a piece of feces. The mother's gain of weight in pregnancy is often interpreted by children as an oral intake analogous to food intake that will be terminated in anal elimination of the newborn baby.

Abraham (1924) suggested subdividing the anal stage into the *anal-expulsive* and the *anal-retentive* stages (table 6.1). In the early, expulsive phase the child does not care for the external object, and enjoys the sadistic expulsion of feces. Folklore and slang bear witness to these anal-aggressive tendencies, which are kept alive often in teenagers and adults. At the late anal stage, the anal-retentive, the child may develop affection for the feces, which become his love object. He may try to keep and to preserve them. Feces are the first possession that the child parts with out of love for the person who cares for him. Feces are the prototype of a gift, and subsequently of gold and money. On the other hand, feces symbolize babies. Often the penis is regarded by children as analogous to the column of feces that fills the mucous tube of the bowel. The anal-retentive phase is considered the source of tenderness. Freud accepted Abraham's suggestions on this point and elaborated on the concept of tenderness in contradistinction to the oral type of love. *Tenderness* stems from the wish to keep and to preserve the object that gives gratification and to take care of it. Only very small babies do not care, and tend to destroy objects that give them pleasure. As the child grows, he cares more for pleasurable objects, wishing them to stay and to continue to serve as a source of pleasure. This new attitude to objects starts with his consideration for breast or bottle. Then it grows into a consideration for his mother, whom he wishes to keep and preserve as a source for a future and continuous flow of gratification. Toward the end of the anal stage, as Abraham found, "retention pleasure" outweighs "elimination pleasure." Later on, as the object cathexis grows, the child cares for his property and his pets and handles them carefully and with tenderness. The sexual instinct becomes *aim-inhibited*. The wish to perpetuate the existence of the love object may take the place of the wish to possess it. In well-adjusted adults, the wish to possess and the wish to preserve are merged in marriage.

The Urethral Phase

The urethral developmental stage is an introductory period to the phallic stage, when genital organs become the main avenue of libido gratification. In both male and female, the urinary tracts are closely related to the genital tracts, and children's sexual fantasies often confuse urine with semen and sexuality with urination.

Urethral eroticism is mainly autoerotic, for one's own body becomes the love object. This eroticism turns toward other objects with fantasies about urinating on them or being urinated on by them.

Urethral eroticism in girls develops sometimes into a retentive pattern, but usually it is aimed at the expulsion of urine and at the pleasure derived from emptying the bladder. The urination itself may be active and aggressive, such as urinating on someone. In boys, urethral eroticism leads to normal and active genital eroticism. In girls it leads to a conflict about their sex role and later on becomes associated with penis envy. The passive nature of urination felt as "let it flow," or loss of control over the bladder, leads in boys to confusion about their sex, and quite often one may find feminine tenderness in men who have been bed wetters in their childhood.

Training for bladder control often leads to conflicts with parents. Delay in assuming this control is often punished by parents in a manner that hurts the child's self-esteem and provokes feelings of *shame*. Bed-wetting children often develop *ambitious* strivings in the struggle against the shame.

The Older Child

The Phallic Phase

Usually at around the age of four the child enters the phallic stage of libidinal development. The name "phallic" is derived from the Greek word *phallos,* which means penis in erection. At this stage, the pleasurable sensations in the genital organs procured by manual stimulation assume the dominant role. The libido is now being "placed" in the genital organs, but it takes time for all sexual excitement to become concentrated in the genitals and discharged by them.

The most important development of the phallic stage is the Oedipus complex. Greek mythology, and later the Greek writer Sophocles, told the story of a young prince, Oedipus, who killed his father, married his mother, and became the king. Oedipus did not know that it was his father whom he had killed and his mother whom he had married, and when he discovered his crime, he punished himself by putting out his eyes. Greek mythology considered the deeds of Oedipus not a result of his malevolence but as an inevitable fate. Freud regarded this legendary story as a symbolic story of the prehistorical development of human society, and, in accordance with Haeckel's biogenetic principle, as reexperienced by children.

According to Freud, something similar to the Oedipus drama happens to the four- or five-year-old boy. He desires to possess his mother physically "in the ways which he has derived from his observations and intuitive surmises of sexual life and tries to seduce her by showing her the male organ of which he is the proud owner." The boy tries to take his father's place; he considers his father a competitor, and develops ambivalent feelings toward him. Although he loves and admires his father, at the same time he hates him intensely and wishes to annihilate him.

At this stage the penis becomes a source of pleasurable sensations. However, in contradistinction to the urethral desire to be fondled, there is a definite need for active push and thrust with the penis. It becomes a most precious source of gratification and pride.

The mother usually notices the masturbatory activities of the little boy, forbids him to play with his penis, and may threaten that unless he stops masturbating she will take away his penis. Often she warns the boy that she will tell the father, and the father will cut the penis off. The little boy is aware of his vulnerability and of his inferiority in relation to the father, whose penis is larger. He is afraid that the father may punish and castrate him. If he has had a chance to notice the difference between male anf female organs, the castration threat becomes very realistic and shocking; he believes that all people originally had a penis but in some cases it was cut off by the omnipotent father. This castration fear is much stronger than the oral fear of being eaten or the anal fear of losing the body content (Freud 1909).

The castration fear forces the boy to abandon his wishes for his mother. He may give up masturbation altogether and develop a passive attitude of a nature ascribed by him to his mother. This passive attitude conceals his increased fear of and hatred for his father, which sometimes develops later into a defiant attitude against all men in authority. Nor does the boy give up his affection for his mother, which often turns into a dependence relationship, into a need to be loved. This attitude, with its strong feminine components and partial identification with the mother, may lead to a submissive attitude toward women.

In some cases, the love for the father is stronger, and the little boy represses his phallic strivings toward his mother. Instead he develops a passive, pregenital sexual desire for his father. This *negative Oedipus complex* may lead to homosexuality.

Sexuality in Females

In boys, the Oedipus complex causes castration fears, which later lead to its resolution; in girls the Oedipus complex proceeds in a reverse order. As soon as the little girl realizes the organ differences between the sexes, she develops the *penis envy* which leads to love for the father. The feminine counterpart of the Oedipus complex is called the *Electra complex* (Freud 1931).

Originally the little girl believes that everyone is built the way she is. As soon as she discovers that some people have penises, she wishes to have one and it occurs to her that she had a penis but lost it. "She begins by making vain attempts to do the same as boys and later, with greater success, makes efforts to compensate herself for the defect—efforts which may lead in the end to a normal feminine attitude." Often she masturbates, using the clitoris as a penis-substitute. If she clings to her wish to have a penis, she may develop masculine tendencies and become domineering and aggressive, and sometimes a homosexual. However, things may develop in another direction. She may become hostile to her mother because her mother did not give her a penis or took it away from her, and because the mother possesses the father. Typical for the Electra complex is the girl's wish to annihilate her mother and to possess her father's penis. Love for the mother turns into hate; and the girl reacts to the loss of this love object by *identifying* herself with it. She wishes her mother's role, and instead of having a penis desires to have a baby, which is a penis substitute. Her role now becomes passive, receptive; the road is paved for normal feminine sexuality.

The sexuality of the girl at this stage is focused in the clitoris. Clitoral masturbation is typical for this age, and it is sometimes accompanied by masculine fantasies in which the clitoris plays the role of a penis. In the case of a negative Electra complex, a girl may dream about taking the role of the father and inserting her clitoris into the mother's vagina and having a baby with her.

In normal cases, the love for the father leads to giving up the wish for a penis, identification with the mother, acceptance of the feminine-receptive role, and wish to have (to "incorporate") a baby.

The differences between masculine and feminine traits are described by Freud as follows:

> When you say "masculine," you mean as a rule "active" and when you say "feminine" you mean "passive." . . . The male sexual cell is active and mobile; it seeks out the female one, while the latter, the ovum, is stationary, and waits passively. This behavior of the elementary organisms of sex is more or less a model of the behavior of the individuals of each sex in sexual intercourse. The male pursues the female for the purpose of sexual union, seizes her and pushes his way into her. (Freud 1933:156)

But, Freud remarked, in some animals the female is the stronger and more aggressive party. Even the function of care for the infant is not always strictly "feminine"; the function of feeding, caring for, and protecting the infant usually performed by mothers is not a passive behavior.

Sex Differences

The best method for studying the differences between the sexes is to observe the way in which men and women develop out of the bisexual disposition of infants. In the earliest stages, little girls are more dependent and docile and learn earlier to control bowels and bladder. In the anal-sadistic stage, girls are no less sadistic than boys. In the phallic phase girls behave as if they were little men, and all their masturbatory activities concentrate around the clitoris, which is a penis equivalent at this stage.

In the pre-Oedipal stage girls are attached to their mothers just as boys are, with all the ambivalent feelings of love and hate of the oral, anal, and phallic stages. The turning away from the mother in the post-Oedipal phases may develop into a bitter resentment against her. The main reason is that the girl holds her mother responsible for her lack of a penis. The boy does not have to change his love object, which is always the mother, nor the erotogenic zone, which is the penis. The girl, on the contrary, has to change both. To become a woman she has to substitute father for mother as a love object and vagina for clitoris as the erotogenic zone.

The difficulties and hazards of the phallic phase for girls are summarized by Freud as follows:

> The discovery of her castration is a turning-point in the life of the girl. Three lines of development diverge from it; one leads to sexual inhibition or to neurosis, the second to a modification of character in the sense of masculinity complex, and the third to normal feminity. The

> little girl . . . finds her enjoyment of phallic sexuality spoilt by the influence of penis envy. . . . She gives up the masturbatory satisfaction which she obtained from her clitoris, repudiates her love towards her mother, and at the same time often represses a good deal of her sexual impulses in general. (Freud 1933:172)

While in boys the Oedipus complex is terminated by castration, in girls it is created by castration.

Freud found that many women take their fathers as models for their choice of a husband or assign their father's place to them, but in married life, they repeat with their husbands their bad relationships with their mothers. Often the mother-relation was actually the one upon which the father-relation was built. The original basis emerges from repression under the pressures of married life. For these women, development to womanhood has consisted mainly in transferring affective ties from the mother to the father object.

> With many women we have the impression that the period of their maturity is entirely taken up with conflicts with their husbands, just as they spend their youth in conflicts with their mothers. . . . Childish love knows no bounds, it demands exclusive possession, is satisfied with nothing less than all . . . It has no real aim; it is incapable of complete satisfaction and this is the principal reason why it is doomed to end in disappointment and to give place to a hostile attitude. (Freud 1931:231)

The Latency Phase

The child is usually protected from external dangers by his parents, and the loss of parental love means the loss of security. The fear of loss of support, combined with the castration fear and reinforced by primeval sources, forces the child to give up his Oedipal incestuous wishes. He "introjects" the parental prohibitions and usually identifies himself with the parent of the same sex. This identification with parental wishes and acceptance of parental standards leads to the establishment of conscience and superego.

At the next developmental stage, the *latency period*, the Oedipal incestuous and aggressive feelings are repressed and forgotten. Part of the instinctual forces is put behind the parental prohibitions and used as anti-instinctual forces. The "internalized" parental prohibitions, which form the superego, threaten the child with severe punishment and keep under severe control the repressed Oedipal cravings. The child identifies himself with the threatening parental figure, which is in normal cases the parent of the same sex. The sexual interest of the child subsides considerably, mainly through inhibitions and sublimations. His love for his parents becomes desexualized and aim-inhibited. Although the sexual-sensual elements are preserved in the unconscious, the sensual goals of his sexuality become inhibited and the feelings toward the parents are tender rather than passionate.

As a result of the inhibition of the Oedipal cravings children give up interest in persons of the opposite sex. During the latency period, usually between the ages of six and eleven, boys play with boys, girls play with girls. The children tend to associate and identify with the parent, adults, and peers of the same sex, and to develop interests that increase their identification with and feeling of belonging to their own sex (Wolman and Money 1980).

Puberty

In order to become psychologically and socially adjusted, the boy must direct his libido away from his mother to another love object, and he must resolve the conflict with his father. This development takes place, as a rule, in the teenage years. The rapid physiological changes of this period bring to full-scale development the sexual urge, and lead to new interpersonal relationships. The genitalia become the main erotogenic zone and the desire for heterosexual contacts becomes dominant. Identification with the proper sex was accomplished in the latency period; now the adolescent strives toward action patterned upon the actions of the parent of the same sex.

In most cases, the sensual and the "tender" undercurrents of love unite at puberty. The adolescent learns to combine the uninhibited and bursting sensual urge with deeply inhibited feelings of care, tenderness, and consideration for his love object. He grows into adulthood and becomes gradually more ready for marriage.

Often the unification of these two elements does not take place or is retarded or partially unsuccessful. In such cases the adolescent and later the adult keeps separate his tender feelings and admiration for women who do not arouse him sexually, and becomes aroused and potent with women for whom he has neither respect nor any tender feelings. Maturation in such cases is far from complete; the tender and aim-inhibited attachment to the mother, normal for the latency period, was not overcome in the teenage years and continues to act as a disturbing factor in adulthood.

In some cases, when the unresolved Oedipus complex remains in a highly intense fixation, the adolescent, instead of giving up the desire for his mother, identifies himself with her. The renounced love object becomes introjected into the ego, and severe emotional disorder may develop.

Fixation and Regression

The development of libido is a biological process, and the consecutive stages represent some innate tendencies to proceed in accordance with the laws of development. The idea of developmental stages was borrowed by Freud from biology, and especially from Darwin and Haeckel. The concept of gradual development as a natural process was strongly supported later by independently conducted observations. Gesell in America and Piaget in Switzerland developed theories of child development independently of Freud. Gesell established the *structure-function* principle, which implies that a certain pattern of activity may not begin until the organism is ready for it. Piaget went even further in suggesting definite developmental stages, some of them strikingly similar to those proposed by Freud.

Some critics have accused Freud of being too "biological"-minded and not "sociological"-minded enough. However, biology has never overlooked environmental factors and ecological considerations; nor was Freud blind to social influences on libido development. Biology and sociology are not opposites, and environment seems to be the determining factor in variations in the development of species.

"Owing to the general tendency to variation in biological processes it must necessarily happen that not all these preparatory phases will be passed through and completely outgrown with the same degree of success; some parts of the function will be permanently arrested," wrote Freud. The usual course of development can be "disturbed and altered by current impressions from without" (Freud 1916).

These variations in libido development are caused by environmental factors. Parts of the libido or of its component impulses sometimes become arrested at an early phase of development. This arrest was called by Freud *fixation*. The fact that some portions of the libido became fixated increases the danger that in a later stage, when facing obstacles the libido may *regress* to those fixations.

Actually, libido development is never perfect and rarely proceeds smoothly from one stage to the next. Often fixations are formed; often regressions take place. While the main body of the libido progresses from one stage to another, some units of it may become fixated. If, owing to unfavorable circumstances, the fixations are substantial, the main body of the libido becomes weakened and less capable of overcoming the external obstacles to development. Neurotic disturbances result from regression to an early developmental stage of the libido.

There are two types of regressions of the libido. The first type takes place when the libido returns to its early love objects. Obviously this regression revives the incestuous wishes. The other type takes place when the entire libido falls back to an early developmental stage.

Frustration may cause regression. However, the sexual urge seems to be much more flexible than hunger or thirst, and there are many ways to endure sexual frustration. Libido can be displaced; its objects can be easily changed; if one component of sexuality is frustrated, another may be satisfied.

Another way of handling sexual frustration is *sublimation*. Sublimation is a diversion of a part of the sexual energy into nonsexual activities. Freud ascribed creative art to sublimation of libidinal energies and their being put at the disposal of the creative talent.

It would be impossible to sublimate the total amount of libido, and there are some definite limitations as to how much sexual deprivation one may take. Sublimation is not always possible, and it discharges only a part of the libidinal energies.

The danger of regression under stress depends upon the strength of fixation and on the amount and duration of stress to which one is exposed. The fixation of libido is the internal, predisposing factor, while frustration is the external, environmental, and experimental factor. The development of the libido with all its unavoidable fixations represents the *predisposing* factor. This is the *sexual constitution* of the individual. On the other hand, the events and frustrations experienced by the individual are a powerful factor in causation of regression and neurosis. The relative importance of constitution and constellation or of fixations and frustrations greatly varies.

In 1931 Freud published a short paper on libidinal types. In this paper he distinguished three libidinal types, namely the *erotic*, *obsessive*, and *narcissistic*.

> The erotic type is easily characterized. Erotics are persons whose main interest—the relatively largest amount of their libido—is focused on love. Loving, but above all being loved, is for them the most important thing in life. They are governed by the dread of loss of love, and this makes them peculiarly dependent on those who may withhold their love from them. Even in its pure form this type is a very common one. Variations occur according as it is blended with another type and as the element of aggression in it is strong or weak. From the social and cultural standpoint this type represents the elementary instinctual claims of the id, to which the other psychical agencies have become docile.

> The second type is that which I have termed the *obsessional*—a name which may at first seem rather strange; its distinctive characteristic is the supremacy exercised by the superego, which is segregated from the ego with great accompanying tension. Persons of this type are governed by anxiety of conscience instead of by the dread of losing love; they exhibit, we might say, an inner instead of outer dependence; they develop a high degree of self-reliance, and from the social standpoint they are the true upholders of civilization, for the most part in a conservative spirit.
>
> The characteristics of the third type, justly called the *narcissistic*, are in the main negatively described. There is no tension between ego and superego—indeed, starting from this type one would hardly have arrived at the notion of a superego; there is no preponderance of erotic needs; the main interest is focused on self-preservation; the type is independent and not easily overawed. The ego has a considerable amount of aggression available, one manifestation of this being a proneness to activity; where love is in question, loving is preferred to being loved. People of this type impress others as being "personalities"; it is on them that their fellow-men are specially likely to lean; they readily assume the role of the leader, give a fresh stimulus to cultural development or break down existing conditions. (1931:248–249)

Cathexis

Whenever Freud was talking about object relations, he applied the concept of cathexis. *Cathexis,* a term borrowed from static electricity, means charge or investment of some amount of energy. Freud, always being faithful to the principles of monism, assumed that mental energy is a derivative and continuation of the biochemical energy of the living organisms, which is, in turn, part and parcel of the universal energy of electric particles. Mental energy is either a creative libido, activated by Eros, the instinctual drive of life, or destructive energy (often called destrudo), activated by Thanatos, the instinctual drive of death. The totality of mental balances led to the formulation of the psychoanalytic concept of character.

Character

The permanent and consistent patterns of overt and covert behavior combined with the totality of ways in which an individual faces frustrations, overcomes hardships, and relates to others is called *character.* Freud's writings imply that ego and character are two different, though closely related terms. Ego is a "structural" term and is defined as part of the mental apparatus. On the other hand, Freud used *character* as a more descriptive term, dealing with the entire behavior of an individual. Character encompasses the history of an individual's libidinal development, his fixations and regressions, defense mechanisms and adjustments, and the totality of interactions between the individual and his environment. For Freud, character represents the quality or the "how" of personality. Each individual personality contains diverse elements, and how these work together, how the individual functions as a whole, and in what ways he differs from other individuals, all this is his character.

In the complex Freudian framework, the interdependence of the various mental forces and agencies is of primary importance. The strength of the character is determined by the relative strength and position of the id, ego, and superego. Of course, the development of the ego and superego depends upon environmental influences. Eventually, the ego is able to deal with the instinctual impulses either in a *cathectic* or in an *anticathectic* manner. In well-adjusted adults the instinctual demands find proper outlets; the ego cathects them. Instinctual energies are cathected in the natural functions of the organism and in the objects of these functions. This type of behavior, typifying the *cathectic* character, perpetuates the instinctual demands and finds avenues of gratification in which the demands of the id, the ego, the superego, and of the external world are brought to a rational harmony.

An individual is said to have a cathectic character if he handles the instinctual demands in a rational way. Impulses that jeopardize inner harmony become either successfully repressed or sublimated (i.e., channeled into new and more acceptable areas). The Oedipal situation, for example, lends itself to a cathected solution. Under normal circumstances, the child gives up the love object and the sexual and destructive aims of his Oedipal wishes, and the energies related to these wishes are not blocked but are cathected in substitute aims and objects. The libidinal energy continues to flow, but it becomes desexualized. Part of this energy may become sublimated if the ego has sufficient strength and if there is a good chance for identification with the parents. Adequate sublimation of impulses reduces the tension, and what is left of the unacceptable instinctual aims and objects can be successfully repressed.

When the ego is unable to act in a cathected way, a reactive or defensive anticathectic type of behavior develops. In the *reactive* type of character, the ego cannot handle the unconscious impulses and it wards them off by anticathexis. However, this involves a substantial expenditure of energy that must affect the individual's mental economy and reduce his efficiency. In this type of behavior, repeated anticathexes are required to block the unacceptable impulses, and anxiety accompanies the conflict between the ego and the repressed impulses, limiting the ability of the individual to obtain full satisfaction in life.

The main defense mechanisms used by the reactive character type are reaction formations and phobias. Both of these mechanisms develop into habitual behavioral patterns or character traits, so that some reactive character types seem to be frigid, as if they had a "feeling phobia" or a fear of letting themselves experience the usual human emotions, while others seem to be hyperemotional as a reaction formation against emotions. Still others develop reaction formations against reaction formations.

Naturally, the development of an individual's character depends on his life history. Freud and Abraham (1924) suggested the following character types related to the developmental stages and possible points of fixation. For example, oral fixation results in an *oral character.* Fixation at this stage is usually a result of either abundant or insufficient oral satisfaction. When oral satisfaction is abundant, the oral character develops an overdependent but rather optimistic attitude. Such persons believe that the world owes them a living and will supply it. Oral deprivation, on the other hand, usually leads to depressive, self-aggressive, and hostile attitudes. In either case, these oral individuals expect that their narcissistic needs will be met by the environment, and they are demanding, self-centered, and request others to take care of them. An oral-erotic fixation makes one a compulsive eater, drinker, smoker, and talker.

As stated earlier, Abraham suggested the division of the oral phase into the early-passive and late-aggressive substages. An oral-passive fixation is related to sucking in the first months of life. As adults, such individuals are passive and overdependent, easily disappointed, and unable to take frustrations. They seem to believe that the world owes them tender maternal love and care, and in their optimistic daydreaming, they hope to receive whatever they wish. The traits of the oral-aggressive type, on the other hand, are related to frustrations in nursing and to biting. This type of individual is critical, hostile, negativistic, and over-demanding, and tends to destroy his love objects. He too seems to believe that the world owes him support and affection, but that for certain reasons, it refuses to fulfill this debt. He is disagreeable and offensive, as if trying to force the world to give him the love and affection that he feels is owed to him.

As we have seen above, Abraham also suggested a division of the anal stage into the anal-expulsive and the anal-retentive substages (Freud 1908). Orderliness, parsimony, and obstinacy are the three main *anal character* traits. Exaggerated and overly harsh toilet training may lead to compulsive behavior in washing and hyperconcern for personal cleanliness, neatness, and orderliness in work and household duties. The excessive parsimony that develops during the anal-retentive phase is expressive of the tendency to retain what one possesses, and feces is the prototype of all transferable possessions.

The *urethral character* type is described as ambitious, impatient, and envious, yet urethrality also includes the tendency toward passivity. Often this type of personality has been punished for enuresis by being put to shame. As a result, the bed wetter would like to hide what has happened and avoid shame. As an adult he developed the burning ambition not to be ashamed any more, and a feeling of envy also develops for all those who have been successful and have avoided humiliation. The urethral type, however, lacks the necessary persistence and aggressiveness, and can only hope, in an oral fashion, that a magic wand will help him to obtain success and glory.

The *phallic character* type develops as a reaction formation to castration fear. Self-assurance, boastfulness, and aggressiveness are the main phallic personality traits, combined with a narcissistic self-love, vanity, and sensitivity. This exhibitionistic and overtly aggressive behavior is actually a reaction formation to castration fear. The phallic character is torn by ambivalent feelings of needing love and fearing it, of courage and timidity, and he tries to overcompensate for his inner fear of castration and doubts concerning his own masculinity by being overtly aggressive and boisterous. But this bold and ostensibly daring behavior serves as a façade behind which insecurity, inner doubts, and anxiety are hidden.

In girls, on the other hand, if penis envy is not resolved or sublimated, a reactive anticathectic character may develop. This unresolved penis envy may be channeled into the assumption of a masculine role or into vindictive feminine behavior displaying the tendency to be humiliated by, and in revenge, to humiliate men.

The *genital character* is usually able to attain normal adulthood. The supremacy of the genital zone over other erotogenic areas, the subordination of all genital aims to the normal heterosexual urge, and a rational choice of a heterosexual love object are the main elements of the genital character. In maturity, the genital character accepts his or her psychosexual role as a husband-father or wife-mother.

A successful sublimation of libidinal and aggressive impulses depends on how much excitation was already discharged. When too much instinctual energy is blocked, reaction formation may take place. In a mature, genital character type sexual and aggressive energy find adequate and socially acceptable outlets. In well-adjusted individuals the diversified instinctual impulses are coordinated, the Oedipus complex is resolved, and reasonable harmony is established between the various elements of personality.

Neurosis

Neurosis is therefore a quantitative *disharmony* between the various strata, forces, and parts of personality. All mental disorders are a personality trouble caused by a failure to grow up and to become a mature, emotionally balanced person. As early as 1900 Freud wrote that "regression plays a no less important part in the theory of the formation of neurotic symptoms than it does in that of dreams" (1900:54).

Originally Freud divided all neuroses into organic, "actual" neuroses and the psychogenic "psychoneuroses." Psychoneuroses are caused by traumatic experiences in childhood.

> The neurotic is incapable of enjoyment or of achievement—the first because his libido is attached to no real object, the last because so much of the energy which would otherwise be at his disposal is expended in maintaining the libido under repression, and in warding off its attempts to assert itself. He would be well if the conflict between his ego and his libido came to an end, and if his ego again had the libido at its disposal. (Freud 1916:194–195)

Every neurosis offers, as it were, a certain gain. Some neurotic symptoms offer a primary gain, that is, an alleviation of the state of anxiety. Anxiety, being a state of tension between the ego and the id or the ego and the superego, is painfully experienced as apprehension, as feelings of inadequacy, fatigue, and depression—general and often diffuse feelings that accompany almost every neurosis.

Neurotic symptoms are a morbid way of escape from the painful anxiety state. A splitting headache that imitates physical illness is one way out; it is easier to assume that one is physically ill than to be torn by inner conflict that demands making a decision. The physical symptom, called by Freud conversion, may alleviate the patient's guilt feeling and justify his procrastination. In such a case this conversion or psychosomatic symptom serves as a primary gain; it is, indeed, an escape from the frying pan into the fire.

Sometimes neurotic symptoms procure *secondary gain* that is helpful in avoiding the condemnation of others; in many cases—splitting headaches, loss of memory, morose moods, or self-inflicted pain—their aim is the winning of sympathy. All mental disorder representing regression into an early developmental stage, secondary gain symptoms can provide a sort of escape into illness, a means of avoiding responsibility.

Freud believed that the time and nature of fixation determine the nature of a given mental disorder. Accordingly, Abraham in his *Short Study of the Development of Libido . . .* (1924) suggested a timetable of libido development and pathogenesis.

BIBLIOGRAPHY

Abraham, K. 1911. Notes on the psycho-analytical investigation and treatment of manic-depressive insanity and allied conditions. *Selected Papers on Psychoanalysis*, pp. 137–156. London: Hogarth, 1949.

—— 1924. A short study of the development of the libido viewed in light of mental disorders. Ibid., pp. 418–501.

Fenichel, O. 1945. *The Psychoanalytic Theory of Neurosis*. New York: Norton.

Freud, S. Standard Edition. James Strachey, ed. 24 vols. London: Hogarth, 1962; New York: Macmillan, 1962.

—— 1900. *The Interpretation of Dreams*. Standard Edition, vols. 4 and 5.

—— 1905. *Three Essays on the Theory of Sexuality*. Standard Edition, 7:125–243.

—— 1908. *Character and Anal Eroticism*. Standard Edition, 8:167–176.

—— 1909. *Analysis of a Phobia in a Five-Year-Old Boy*. Standard Edition, 10:5–149.

—— 1916. *Introductory Lectures on Psychoanalysis*. Standard Edition, 15:15–242.

—— 1931a. *Libidinal Types*. Standard Edition, vol. 21:217–220.

—— 1931b. *Female Sexuality*. Standard Edition, 21:225–243.

—— 1933. *New Introductory Lectures on Psychoanalysis*. Standard Edition, 22:3–184.

Wolman, B. B. and J. Money, eds. 1980. *Handbook of Human Sexuality*. Englewood Cliffs, N.J.: Prentice Hall.

[SEVEN]
Personality Structure

FREUD'S major work, *The Interpretation of Dreams,* carried the notion that the forces which repress unacceptable, unconscious wishes are themselves conscious or unconscious. The repressing censor was believed never to be unconscious. However, Freud soon discovered that the same forces that repressed unacceptable desires were not necessarily conscious; his patients were unaware of their repressed desires, and it took a great deal of ingenuity and effort on the part of Freud to unravel the hidden thoughts and wishes. In 1900 Freud described how the conscious perceptory system (Pcpt) controls the unconscious:

> Excitatory material flows into the Cs sense organs from two directions: from the Pcpt systems, whose excitation, determined by qualities, is probably submitted to a fresh revision before it becomes a conscious sensation, and from the interior of the apparatus itself, whose quantitative processes are felt qualitatively in the pleasure-unpleasure series when, subject to certain modifications, they make their way to consciousness. (1900;615–616)

But this process, as Freud later found, was more complex. The censor itself acted in a hidden manner and the individual often was unaware what was repressed. Apparently, the controlling and repressing force could belong to any of the three mental strata, conscious, preconscious, and unconscious. This discovery forced Freud to develop a new model of personality in which the *topographic division* (mental strata) would be no more than one of the aspects of mental functioning.

In 1923 Freud introduced what was for him a somewhat new model of personality structure based on economic, topographic, and structural considerations. This model dealt with the distribution, balance, and mutual interdependence of the two instinctual forces, Eros and Thanatos, and the energies at their disposal, the libido and the destructive energy—or the *economy* of the mind; it also dealt with the three mental provinces, the unconscious, preconscious, and conscious—or the *topography* of the mind; and it dealt with the three mechanisms of personality, the *id*, the *ego*, and the *superego*—or the structure of the mind (Sulloway 1983; Chs. 6 and 10).

The Id

Freud's new personality model introduced a division of personality into three parts, the id, the ego, and the superego. The neonate has id only; the ego and superego develop later in life. Whatever is inherited or fixed at birth finds its first mental expression in the id. The id actually becomes the link between somatic and mental processes. It is "somewhere in direct contact with somatic processes, and takes over from them instinctual needs and gives them mental expression, but we cannot say in what substratum this contact is made" (Freud 1933; 104).

The id expresses only the true purpose of the individual organism's life, namely, the immediate satisfaction of its innate needs.

The id has no organization and no unified will, only an impulsion to obtain satisfaction for the instinctual needs, in accordance with the pleasure principle (Freud 1932).

The cathexes or energy discharges of instinctual impulses in the id are "unbound" and highly mobile, in contradistinction to the "bound" cathexes of the ego and superego. As a result of this unstable organization of excitatory processes, the id may "displace" or "condense" instinctual impulses in its attempt to insure immediate gratification. The id transfers cathexes from an idea or process that is unable to gain access to an available outlet to another idea or process that is free to discharge energy. This transfer of energy is called displacement. Sometimes an idea or process acquires the whole cathexis of several other ideas or processes; this mechanism is called condensation. The id, being a "cauldron of seething excitement" is incapable of logical reasoning, which requires "bound" energy. Contradictory instinctual wishes coexist, side by side, in the id, and they may displace or condense their cathexes onto one another. The instinctual elements of love-hate motivated by Eros and Thanatos respectively may fuse or merge their energies in the id. The unbound mental processes of the id have no conception of time. Infantile wishes repressed into the id by the ego may forever strive for gratification; they may manifest themselves in dreams, symptom formation, and parapraxes.

> The laws of logic—above all, the law of contradiction—do not hold for processes in the id. Contradictory impulses exist side by side without neutralizing each other or drawing apart; at most they combine in compromise formations under the overpowering economic pressure towards discharging their energy. (Freud 1933:104)

At the id level of functioning, behavior is always motivated by a desire to avoid or, at least, to decrease physiological and/or psychological tensions. Tension or pain is subjectively perceived whenever there is an increase in the quantity of excitation present in the somatic-psychic life. Relief or pleasure is experienced whenever there is a diminution of excitation. The id is ruled by the tendency of the psychic apparatus to keep the quantity of tension or excitation to a minimum or constant level called the constancy principle.

The restoration of equilibrium is perceived by the organism as relief or pleasure. "In the psychoanalytical theory of the mind," Freud wrote, "we take it for granted that the course of mental process is automatically regulated by the pleasure principle: that is to say that we believe that any given process originates in an unpleasant state of tension and thereupon determines for itself such a path that its ultimate issue coincides with the relaxation of this tension, i.e., with the avoidance of 'pain' " (Freud 1920:62).

The id is the somatic source of life and was described by Freud as the "reservoir of instinctual drives" which are continually striving for immediate satisfaction. Instincts are the psychological concomitants of biochemical processes. "An instinct appears to us a borderland concept between the mental and the physical, being both the mental representative of the stimuli emanating from within the organism and penetrating to the mind, and at the same time a measure of the demand made upon the energy of the latter in consequence of its connection with the body" (Freud 1915a: 127).

Pleasure Principle

The id knows no precautions to ensure survival, and in fact, an immediate and unconditioned gratification of instinctual demands could lead to a dangerous clash with the external world and to the death of the organism.

The entire mental energy is originally stored in the id. This energy is put at the disposal of the organic instincts, which are composed of fusions of two primal forces, Eros and destructiveness. The only action of all the instincts, related to the various somatic organs, is satisfaction or immediate discharge of energy to remove tension and bring relief. This is what Freud meant by the *pleasure principle;* it is the urge for an immediate discharge of energy that brings an immediate relief and pleasure.

The id is completely unconscious, and therefore, at the beginning of life, everything is unconscious. Due to the constant stimulation by the outside world, part of the unconscious material of the id develops into preconscious material and the ego emerges. Not all of the preconscious material remains the permanent property of the ego, however. Part of it is repressed and lost again to the unconscious. In adults, the id's unconscious material is composed both of the original, unaltered, and almost inaccessible unconscious nucleus and of a relatively younger and more easily accessible portion that has been repressed by the ego and thrown back into the id.

The mental processes in the id, called by Freud the *primary,* greatly differ from the overt sensation and perception. Although the id is cut off from the external world, it does perform perceptory functions in its own interior. The mere fact that the id is governed by the pleasure principle and acts always in the direction of procuring pleasure and avoiding displeasure points up to the fact that the id is capable of perception. The self-directed perceptions and coenesthetic feelings disclose the economy of inner tensions and the balance of the mental apparatus. When this economy or the equilibrium of the mental apparatus is disturbed, the instinctual forces react in the striving for immediate discharge of energy. But the id, blindly obeying the pleasure principle, knows no values, no right or wrong, no moral standards, and no considerations for other people. Freud called it a "cauldron of seething excitement," a collection of "instinctual cathexes seeking discharge." The energy in the id is unbound, fluid, capable of quick discharge.

Another characteristic of the id is its unchangeability. "Conative impulses which have never gotten beyond the id, and even impressions which have been pushed down into the id by repressions, are virtually immortal and are preserved for whole decades as though they had only recently occurred" (Freud 1933:104). So we see that the repressed remains unaltered by the passage of time. In dreams, long-forgotten infantile memories come back, and in mental disorders, early childhood experiences play a decisive role.

It remains certain, Freud wrote in the last summary of his theory,

> that self-perceptions—coenesthetic feelings and feelings of pleasure-unpleasure—govern events in the id with despotic force. The id obeys the inexorable pleasure principle. But not the id alone. It seems as though

> the activity of the other agencies of the mind are able only to modify the pleasure but not to nullify it; and it remains a question of the greatest theoretical importance, and one that has not yet been answered, when and how it is ever possible for the pleasure principle to be overcome. (1938:109)

When the other mental agencies, the ego and the superego develop, their energies are borrowed or derived from the id. It cannot be overemphasized that, in the Freudian system, the only source of mental energies both of Eros and Thanatos is the id, and parts of its energies become invested in the higher mental agencies.

Ego Emergence

Although the neonate is exposed to stimuli, he is unable to perceive clearly what is going on, or to move voluntarily, or to master the excitations created by inner and outer stimuli. These three functions—perception, voluntary mobility, and control of tensions—eventually become the functions of the emerging ego. At birth, however, the newborn has no ego, and he is exposed to excitations which he cannot master. He can only become dimly aware of them, and he feels uncomfortable and unhappy. Hunger, thirst, cold, noise, and other tension-producing stimuli flood his mental apparatus and produce a state of anxiety.

Of course, the infant's first reaction or tendency toward reaction is concerned with removing the disturbing stimuli. The infant himself however, is unable to do this; he is helpless and cannot survive without being taken care of. Help must come from the external environment.

For Freud, the neonate's mental apparatus resembles a body floating in water. Only the surface is exposed to the outer world, and it receives external stimuli and discharges motions. At first, the entire apparatus is id, but under the influence of environmental forces acting on the surface of the id, the surface undergoes substantial changes and gradually develops into a separate part of the mental apparatus called ego. It must be emphasized that the influence of the environment produces changes in the external part of the id only. The unconscious material of the external id becomes transformed into the preconscious ego, in which the primary mental processes give ground to the emerging secondary processes.

As the child grows, his ego gradually becomes more capable of protecting the organism against threats coming from within and without. The emerging ego is able gradually to draw the line between self and the external world. The undeveloped, primitive ego can only *introject* objects, making them part of the self, or *project,* ascribing parts of itself to the outer world.

In general, a child accepts pleasant stimuli as being part of the self and ascribes unpleasant ones to the outer world. Freud stressed that a reasonable amount of frustration is necessary for normal development. Obviously, a child whose every wish comes true can never develop a proper distinction between himself and the outer world. The wall that doesn't move away when the child pushes it enables the child to draw the line between reality and wishful thinking.

The id's impulsive reactions based on pleasure and constancy principles do not insure survival. In reaction to external dangers, the id develops a sort of protective shell called the ego (Freud 1923).

The ego slowly develops out of the id in its attempt to cope with the demands of the external world. The need for self-preservation dictates the ego's modification of overt behavior and control over the id's primary processes. The ego's actions are guided by the reality principle. The aims of all behavior are to secure instinctual gratification, the ego refuses to take unnecessary risks, and it defends the individual's life rather than denounce pleasure, if in the process of procuring it, they could be seriously hurt or killed by themselves or by others.

> The first case of such a check on the pleasure principle is perfectly familiar to us in the regularity of its occurrence. We know that the pleasure principle is adjusted to a primary mode of operation on the part of the psychic apparatus, and that for the preservation of the organism amid the difficulties of the external world it is *ab initio* useless and indeed extremely dangerous. Under the instinct of the ego for self-preservation, it is replaced by the reality principle, which, without giving up the intention of ultimately attaining pleasure, yet demands and enforces the postponement of satisfaction, the renunciation of manifold possibilities of it, and the temporary endurance of "pain" on the long and circuitous road to pleasure. (Freud 1920:13)

Gradually the ego assumes control over the organism's cognitive and motor apparatus. In order to decide which instinctual impulses can be safely gratified the ego must be realistic and correctly perceive the demands and restrictions of the environment.

> On behalf of the id, he ego controls the path of access to motility, but it interpolates between desire and action the procrastinating factor of thought, during which it makes use of the residues of experience stored up in memory. In this way it dethrones the pleasure principle, which exerts undisputed sway over the processes in the id, and substitutes for

> it the reality principle, which promises greater security and greater success. . . . The ego advances from the function of perceiving instincts to that of controlling them. (*Ibid.*)

Reality Principle

The chief task of the ego is self-preservation:

> by avoiding excessive stimuli (through flight), by dealing with moderate stimuli (through adaptation) and, finally by learning to bring about appropriate modifications in the external world to its own advantage (through activity). In regard to internal events and in relation to the id, the ego performs its task by gaining control over the demands of the instincts by deciding whether they shall be allowed to obtain satisfaction, by postponing that satisfaction to times and circumstances favorable in the external world, or by suppressing their excitations completely. Its activities are governed by consideration of the tensions produced by stimuli present within it or introduced into it. The raising of these tensions is in general felt as *unpleasure* and their lowering as *pleasure* . . . The ego pursues pleasure and seeks to avoid unpleasure. An increase in unpleasure which is expected and foreseen is met by a *signal of anxiety*. (Freud 1938:15)

The id seeks pleasure, come what may, but the ego is concerned with discovering the most favorable and least dangerous method for satisfaction. The id is bound by the necessity of seeking immediate gratification of instinctual demands, while the ego is capable of logical reasoning, of considering causal relations, and of learning by experience.

The ego's main task is self-preservation, but to protect the organism, the ego has to perform several functions. The first is *testing reality*—acting as a constant watchdog, perceiving the outer world and thus making possible avoidance of confusing inner and outer stimuli. For this reason, the ego is said to function in terms of the *reality principle*. The second task is the control of the unconscious impulses coming from the id and the superego, and the third task is the control of the motor apparatus.

Reality Testing. Since the id is cut off from the external world, it is the task of the ego to distinguish between the inner, id-originated stimuli, and the stimuli from without, perceived by sensory apparatuses. The archaic ego and the failing ego of psychotics may not be able to distinguish between inner and outer stimuli and may fall prey to delusions and hallucinations. The mature ego, on the other hand, is capable of *reality testing*, which consists of checking its perception against the outer reality and eliminating from the picture of the outer world any elements that may stem from "inner sources of excitation." The mature ego separates wish from reality thus making possible adjustment to real situations. In sleep, when the ego's reality-testing function is suspended, inner and outer stimuli merge in the world of fantasy and wish fulfillment of dreams.

Infantile perceptual and motor functions are diffuse and intertwined, but gradually the infant's passive experience of excitations is transformed into observation, retention of perception (memory), and motor responses. The growing ego becomes an organized entity with differentiated functions; one of them is the protection of the organism from too many or too strong stimuli.

The ego's perception of the external world begins with a primary, archaic identification with perceived objects. Several functions are involved in this archaic identification. An infant often puts perceived things into his mouth, and in so doing, introjects his first love objects. This kind of *oral introjection* shows that several distinct functions such as perception, motility, and emotionality have not separated from one another. Identification is an effort to master stimuli that are too intense by adjusting one's body to them and imitating them.

According to Freud, the "surface" part of the ego is the perceptual-conscious system. This system "is directed on to the external world, it mediates perceptions of it, and in it is generated while it is functioning, the phenomenon of consciousness. It is the sense organ of the whole apparatus, receptive not only of excitations from without but also of such as proceed from the interior of the mind" (Freud 1933:106).

Anticathexes

The ego must meet the inner demands of both the id and the superego, and the external demands of reality. Simultaneously, the ego also has to preserve its own organization and maintain its autonomy. The economic task of the ego is to reduce the forces that act upon it and to bring to them a degree of harmony and balance. The pressure coming from the id may either disrupt the ego and turn in back into id, or by forcing the ego to disregard reality, expose the entire organism to external dangers.

The ego is forced to expend considerable amounts of energy in anticathexes to keep down the instinctual pressures coming from the id. The superego, on the other hand, may become too powerful and press the ego. The id and the superego may join forces against the ego, so that the hard-pressed ego barely manages to keep its relation to reality. The ego, weakened by the internal conflict, loses its contact with reality and slips down into psychosis.

The newborn does not have the ability to control the impulse for immediate discharge of energy, nor can the archaic ego master excitation. The mature ego, on the other hand, is able to absorb greater amounts of excitation and is able to check panic reactions by anticathexes. The stronger the ego, the greater is its tolerance of frustrating experiences, and its ability to control emotional reactions. The strength of the ego is a good indicator of mental health.

Motility Control

Freud wrote, "We have here formulated the idea that in every individual there is a coherent organization of mental processes which we call his *ego*. The ego includes consciousness and it controls the approaches to motility, i.e., to the discharge of excitations into the external world" (1923:13). One of the main tasks of the ego, then, involves adjustment to the organism's environment, and adjustment means action. The mature ego, which controls the motor apparatus and the cognitive functions, inhibits and regulates the motor functions of the organism. In infancy, external stimuli elicit an immediate and unorganized mass reaction of the entire organism. One of the most important functions of the ego is to "bind" the instinctual impulses and to convert their mobile cathectic energy into a predominantly quiescent, tonic cathexis. An infant's actions are impulsive, while adult behavior is more controlled.

Freud (1921) maintained that the ego postpones motor reaction and abides by the *reality principle*. As mentioned above, the reality principle is not contradictory to the pleasure principle, but is, rather, a development of the latter. The ego pursues pleasure as much as the id does, but the ego is concerned with the avoidance of unpleasure and the prevention of disastrous consequences of the blind pursuit of pleasure. The ego seeks the best ways of procuring maximum pleasure with a minimum of unpleasure. Although the immediate discharge of energy in an act of love or hate may bring temporary relief, its consequences may be disastrous. The ego, therefore, weighs the possible outcomes before responding to a given stimulus. The mature ego can withstand a certain amount of tension, and it discharges energy in a way that guarantees the least risky gratification of instinctual needs.

As the infant learns to walk and talk, his contact with the outer world increases. He is able to reach objects he could not reach before and he meets new people and experiences new situations. The child's narcissistic and self-centered world gradually gives way to the world of objects and people. No longer bound to a passive and hallucinatory imagination and expectation, the child moves around, becomes acquainted with the world, and tries to master it. The development of the speech function enables the child to exchange information witht the outer world and thus enlarges his field of knowledge. He begins to anticipate the reactions of adults, and his growing ego learns to avoid unpleasurable experience.

> [The ego] controls the path of access to motility, but it interpolates between desire and action the procrastinating factor of thought, during which it makes use of the residue of experience stored up in memory. In this way, it dethrones the pleasure principle, which exerts undisputed sway over the processes in the id, and substitutes for it the reality principle, which promises greater security and greater success. (Freud 1933:106)

Anxiety

Freud originally regarded anxiety as being a result of thwarted sexual urges. Unsatisfied libido and undischarged excitation cause anxiety neurosis. To Freud, it seemed as if the unsatisfied libido were directly transformed into anxiety.

Freud wrote in 1895: "The mechanism of anxiety neurosis is based on the division of somatic sexual excitation from the psyche and the resultant abnormal utilization of that energy." The same ideas are expressed several times in Freud's writing. It conveyed simply the conviction that dammed up sexual libido is transmuted into anxiety. In a letter to Fliess, Freud related coitus interruptus to anxiety neurosis.

In Draft A (probably 1892) and Draft B (1893) Freud traced the etiology of anxiety to disturbance in sexual functioning. In draft E (1894) Freud wrote: "It quickly became clear to me that my neurotic patients' anxiety had much to do with sexuality." Particularly, anxiety has been observed in virginal subjects, prudish people, in people who are sexually abstinent out of necessity, in women with no supply of sexual intercourse, in men who practice coitus interruptus, in men who commit intercourse beyond their desire or strength, and finally, in men abstinent for contingent circumstances. Thus anxiety is "an accumulation of physical sexual tension. . . . Anxiety arises from a transformation of the accumulated tension" (Freud 1887–1902:106).

Holt wrote that Freud built his theory "much as F. D. Roosevelt built the executive branch of the government: when something wasn't working very well, he seldom reorganized; he just supplied another agency or concept to do the job" (1965:166–167). Holt's remark conveys a partial truth, for Freud never refused to refute his past ideas.

Both ideas of birth trauma and interrupted coitus as sources of anxiety were predominant in Freud's theory for a long time. When Freud revised his theories, these two concepts were not abandoned in their entirety, but assigned a lesser role in the new theoretical framework.

The monograph *Inhibitions, Symptoms, and Anxiety* was written in 1925 after Rank published his *Birth Trauma* and after Freud has introduced his structural theory in 1923 in *The Ego and the Id*.

Main changes in Freud's theory revolved around his theory of motivation. Freud abandoned in 1914 the ego versus libido theory and merged these two concepts in the theory of narcissism and later added the Eros versus Thanatos dualism in 1920. In this framework anxiety could no longer be related to a simple frustration of the sexual drive.

The structural personality model included the id, the ego, and the superego and presented anxiety as a problem of inner conflict. The struggle of the ego against the onslaughts of the id and the superego has become the main issue. For a while the old concept of thwarted sexuality was delegated to the so-called actual neuroses, but even there it was totally abandoned. The new theory of anxiety has been described by Freud.

Apparently it is the psychological reaction to stress that has to be considered in this connection. The ego's reaction to external threats is fear. And when the ego is exposed to threats from within, that is, from the id or superego, the feeling of anxiety ensues. The term anxiety was used by Freud in more than one connotation. Originally Freud believed that anxiety was the result of blocking of sexual needs. The combination of unsatisfied libido and undischarged excitation was supposed to cause anxiety-neurosis; the thwarted libido was transformed into a state of anxiety.

Having introduced the structural theory in 1923, Freud introduced a new theory of anxiety three years later. The new theory did not discard the old one, but reduced the scope of its meaning to some few special cases. According to the 1926 theory, anxiety originates from the infant's inability to master excitations. The neonate is exposed to more stimulation than he can possibly master. The abundant stimulation is traumatic, and creates the painful feeling of primary anxiety.

Birth trauma is the prototype of all future anxiety states (cf. Rank 1929). Separation from mother is another anxiety-producing experience. Castration fears, guilt feelings, fear of abandonment and rejection are the most frequently experienced anxiety-producing situations. The feeling of being helpless is one of the most frequent symptoms of neurotic disturbances; it is especially typical of traumatic neuroses. Also, the inability to control one's own excitation (whether aggressive or sexual) may create a state of anxiety.

The early theory of anxiety became incorporated in the new and more broadly conceived theory. Since the satisfaction of instinctual demands may create a dangerous situation, the ego must control the instinctual impulses. A strong ego accomplishes this task easily; but a weak ego has to invest more energies in an anticathectic effort to ward off the unconscious impulses.

Our next question will be: How can we picture the process of repression carried out under the influence of anxiety? I think this is what happens: the ego becomes aware that the satisfaction of some nascent instinctual demand would evoke one among the well-remembered danger situations. This instinctual cathexis must, therefore, somehow or other be suppressed, removed, made powerless. We know that the ego succeeds in this task if it is strong, and if it has assimilated the impulse in question into its organization. In the case of repression, however, the impulse is still part of the id, and the ego feels weak. In such a contingency, the ego calls to its aid a technique, which is, at bottom, identical with that of normal thinking. Thinking is an experimental dealing with small quantities of energy, just as a general moves miniature figures about over a map before setting his troops in motion. In this way, the ego anticipates the satisfaction of the questionable impulse, and enables it to reproduce the painful feelings which are attached to the beginning of the dreaded danger-situation. Thereupon the automatic mechanism of the pleasure-pain principle is brought into play and carries through the repression of the dangerous impulse.

Types of Anxiety. Initially, anxiety starts with the infant's inability to master excitations. The infant is unavoidably exposed to powerful stimulations that are beyond his limited coping ability, and these traumatic experiences create in the child the feeling of desperate helplessness. This is the painful, horrible feeling of *primary anxiety*. The first experience of this kind is the birth trauma, in which the main characteristics of anxiety, such as the accelerated action of the heart and lungs, play a decisive role in the child's survival. Later, separation from the mother is anxiety producing.

This state of painful helplessness or anxiety may be revived in later life under trying circumstances. In fact, the feeling of being helplessly flooded is one of the main symptoms in practically all neurotic disturbances, and especially in a traumatic neurosis. The inability to control excitation, whether stemming from sexual or aggressive impulses, creates the state of anxiety, which is always experienced as an inability to master excitation. By this interpretation, the early theory of anxiety becomes a part of the later and more broadly conceived theory, which says that any anxiety can be traced back to situations of external danger. Sooner or later, the ego realized that the satisfaction of some instinctual demands may create one of the well-remembered danger situations. The ego must then inhibit the instinctual wishes. A strong ego accomplishes this task easily, but a weak ego has to invest more energies in a countercathectic effort to ward off the repressed impulse.

According to Freud, there are three types of anxiety-producing situations in childhood, but these can be reduced to one basic type. These are the situations of being left alone, being in the dark, and finding a strange person in place of the mother. All of these situations represent the feeling of loss of the loved person, and anxiety seems to be a reaction to the perception of absence of or separation from the loved object. Anxiety is probably experienced by an infant in birth trauma, in weaning from the mother's breast, and later on in fears of castration.

> If the infant longs for the sight of the mother, it does so surely, only because it already knows from experience that she gratifies all its needs without delay. The situation, which the infant appraises as "danger" and against which it desires reassurance, is therefore, one of not being gratified, of an *increase of tension arising from non-gratification of its needs*—a situation against which it is powerless. (Freud 1926:76)

Anxiety, then, is a sign of the ego's weakness. If the ego is hard pressed by external reality, it develops "reality anxiety". If it is hard pressed by the superego, which creates a feeling of guilt and inferiority, so-called moral anxiety appears. "Neurotic" anxiety develops when the pressures of the id threaten to disrupt the ego.

Objective, or reality, anxiety is a reaction to external danger or a reaction to an anticipated threat. Anxiety preparedness may develop either by re-experiencing the past experiences as a signal of danger and facing the danger by an action of flight or fight, or by re-experiencing the past danger in its totality with all its paralyzing effects leading to an utter failure in counteracting the present danger.

There are three ways that neurotic anxiety is experienced. One is anxiety neurosis, usually felt as some sort of general apprehensiveness, dreadful expectancy, and uneasiness. Anxiety neurosis is usually caused by undischarged excitation, and the unsatisfied libido energy is transformed into anxiety. Another type of neurotic anxiety takes place in hysteria and in some other severe neuroses. Here, the ideas attached to libido become repressed and distorted, and the energy, whether libidinal or destructive, turns into anxiety. In the first neuroses of childhood (i.e., *phobias*), an internal danger or fear of one's own libido or death instinct becomes transformed into a fear of external dangers. In such cases, it is as if the neurotic anxiety has become externalized and is perceived as objective anxiety or fear of an external threat. The neurotic transforms the fear of his own libido (or death instinct), from which there is no flight, into an external danger that can be supposedly warded off by obsessive behavior.

The reaction to pressure which is exercised by the superego is called moral anxiety. It appears as feelings of guilt or shame or a feeling of one's inferiority and inadequacy. The factual threat to an organism does not necessarily produce anxiety. Anxiety can be produced by an imaginary threat, by inner tensions, or by any other factor that is experienced as a threat.

In the final analysis, then, it is not the danger itself that causes anxiety; obviously, the pleasure principle is not identical with self-preservation.

> What is it that is actually dangerous and actually feared in such a danger situation? It is clearly not the objective injury which need have absolutely no importance psychologically, but it is something which is set up in the mind by it. Birth, for example, our prototype for the state of anxiety, can hardly in itself be regarded as an injury, although it may involve a risk of injury. The fundamental thing about birth, as about every danger situation, is that it evokes in mental experience a condition of tense excitation which is felt as pain and which cannot be mastered by discharge. . . . The operation of the pleasure principle does not guarantee us against objective injury but only against a particular injury to our mental economy. . . . The magnitude of the excitation turns an impression into a traumatic factor which paralyses the operation of the pleasure principle and gives significance to the danger situation. (Freud 1933:130–131)

Defense Mechanisms

The term "defense mechanisms" is the name for techniques by which the ego wards off the instinctual demands of the id or the pressures of the superego. With the exception of sublimation, all defense mechanisms indicate an inner conflict.

One of the most common and important defense mechanisms is *rationalization*. In its attempt to mediate between the id and reality, the ego is often forced to ascribe rationality to the irrational demands of the id, to display a "pretended regard" for reality and, with "diplomatic dishonesty," assume that the id accepts reality while it actually distorts it. Rationalizations may be defined basically as a fallacious reasoning which serves to represent overirrational motivation as if it were rational and, at the same time, to protect self-esteem. It is used as a covering up for mistakes, misjudgments, and failures, and it tries to justify behavior by reasons that seem to be rational.

A more serious distortion of reality occurs when the mechanism of *undoing* is utilized. In undoing, the individual seems to believe that he can undo or nullify his former actions that make him feel guilty. It is a kind of magic, said Freud, to "blow away" not only the consequences of an event, but the event itself.

Still another step further in moving away from reality is the mechanism of *denial*. Reality may become so painful and so dangerous that the infantile ego withdraws from any contact with it and refuses to acknowledge its existence. Although memory and perception prevent the unlimited denial of reality, in some pathological cases, the hard-pressed ego gives up reality testing and applies the mechanism of negating the unpleasant experience.

Introjection is the desire to swallow the love object and it is the prototype of any object relations. It represents the primitive and ambivalent attitude in which love and destruction are combined in incorporating the love object and in identification with him. In later life, if the individual is unable to develop more mature object relationships, the hard-pressed ego may set primitive introjection as an instinctual aim and accept identification as the only possible object relationship. The opposite of introjection is *projection*.

Compulsive personalities often use the defense mechanism of *isolation*. By means of this mechanism an emotional content is separated from the idea into which the emotion was cathected, thus separating two parts of the same experience. In cases of "split ego" or "dual personality," for example, part of an experience is kept separate from the rest of one's ego; usually the part is very unpleasant and cannot be accepted by the ego. Isolation usually takes place after a particularly unpleasant experience that the ego cannot accept. The ego is unable to face the pain or humiliation so it stops functioning for a while.

One of the most important defense mechanisms is *reaction formation*. It is related to repression in that, when a wish or desire is repressed, the ego tries to prevent its reappearing. One way to keep down a repressed impulse is to develop a wish opposite to the original and repressed one. An individual who hates his father, for example, and who is very unhappy about it, may develop a ritual of affection directed toward his father.

Successful and normal defenses against objectionable instinctual wishes are called *sublimations*. Freud defined sublimation as a cathexis of instinctual energy into a substitute aim or object or both and a channeling of the instinctual demands into new and desexualized strivings.

Repression

The unconscious exclusion from consciousness of objectionable impulses is called *repression.*

The repression of instinctual impulses and the action undertaken to protect it, called "resistance," requires a constant expenditure of energy by the ego. This anticathectic energy is derived from the id.

The repressed instinctual desire has no *direct* access to consciousness, but it persists in the timeless domain of the id and may exert constant pressure for a direct or disguised gratification. Freud wrote in 1926 as follows:

> An important element in the theory of repression is the view that repression is not an event that occurs once but that it requires a permanent expenditure of energy. If this expenditure of energy were to cease, the repressed impulse, which is being fed all the time from its sources of energy, would seize the next occasion to flow along the channels from which it has been forced aside, and the repression would either fail in its purpose or would have to be repeated an indefinite number of times. Thus it is because instincts are incessant in their nature that the ego has to make its defensive action secure by a permanent expenditure of energy. This action undertaken to protect repression is perceptible as *resistance* in analytic treatment. Resistance presupposes the existence of what I have called an *anticathexis.* The mechanism of repression, in some form or another, is involved in almost all of the defense mechanisms. Objectionable wishes or ideas are usually removed from consciousness and forgotten, the ego pushing them down into the unconscious and acting as if they were now extinct. Once the instinctual impulses are repressed, the ego endeavors to keep them repressed forever. When the repressed material comes close to the surface and tries to come back into consciousness, the defense mechanisms of the ego push it back into the unconscious. The same forces that cause repression keep the repressed material under a close guard and resist its unearthing. (1926:114)

The "Quality" of the Ego

Freud reasoned that in order to perform its great variety of tasks (i.e., learning from experience, sizing up the present situation, foreseeing future consequences, and being capable of control, postponement, and suppression of instinctual demands), the ego must be a highly complicated apparatus. It must control the entire sensory apparatus of perception, master the motor apparatus, and be able to withstand pressures coming from within and without (Blanck 1977).

The great majority of the ego, Freud states, belongs to the "mental province" of the preconscious. However, it has access to consciousness and it can easily become conscious without any effort. A considerable part of the ego is also unconscious, and it is precisely this part that may be described as its nucleus. Even these internal processes of the ego may acquire the quality of consciousness, however, so that the ego can be stimulated from within even more than from without. This is merely Freud's explanation of the fact that perception may stem from within, from memories, and from inside of the body. In its early life, the ego often confuses inner stimulation with stimuli coming from the external world. These confusions are called illusions or hallucinations, and they are said to occur whenever the ego is not fully alert (i.e., in early childhood, in the dreams of adults, or in psychotic states). The mature ego is better able to test reality and can distinguish between inner and outer stimulation, between memory traces associated with verbal residues and actual, external reality.

According to Freud, the ego gets its energy on loan from the id. The instinctual demands of the id allow the investment of part of the id's energy in objects. This object cathexis enables the ego to draw some amounts of energy from the id by means of various devices, one of them being identification. By identifying itself with the cathected objects, the ego "recommends itself to the id in the place of the object and seeks to attract the libido of the id onto itself. . . . In the course of a person's life the ego takes into itself a large number of such precipitates of former object-cathexes" (Freud 1932:108).

The Superego

Freud explained self-criticism, self-reproach, and feelings of inadequacy and guilt in terms of his concept of superego. As early as 1896, Freud noticed that people use socially accepted norms for self-criticism, and that this self-criticism is the inner force that prevents dangerous and morally unacceptable ideas from appearing in one's conscious. Thus, the so-called censor was held responsible for repression and dreamwork.

In 1914, Freud introduced the term *ego-ideal,* which he defined as a "critical faculty" within the ego. In 1921, he described it as the heir to original narcissism, a sort of self-serving legislative organ. The ego-ideal is formed under the influences of the environment and it represents the demands which that environment makes on the ego and which the ego cannot always rise to. It is really the individual's conscience, the critical self-attitude of a part of the ego. It exercises the power of censorship in dreams and it serves as the main force in repressing instinctual wishes.

The superego is said to evolve out of two highly important factors: the biological and the cultural-historical. The differentiation of the superego from the ego is a product both of the development of the individual and of the species. The ego-ideal or superego sets behavioral norms: do this and don't do that. As the ego is essentially the representative of the external world and reality, the superego is the representative of internalized cultural values. The conflicts between the ego and the superego reflect the contrasts between the individual, realistic, here-and-now situation and the cultural-historical heritage.

In Freud's earlier works the term ego-ideal was synonymous with what he later called superego, but in the later works, the ego-ideal was said to represent a part of the superego and its craving for perfection.

Around the age of five, the clash between the instinctual demands of the id and the fear of punishment leads to the formation of the superego. The ego grows and develops from a state of merely perceiving instincts toward actually controlling them; from yielding to instincts toward inhibiting them. In this development a large share is taken by the ego-ideal, which is partly a reaction formation against the instinctual processes of the id.

Actually, the superego develops as a result of the weakness of the infantile ego. At about the age of two the child undergoes toilet training, and the fear of punishment and the need for affection and protection force the child to accept the parental demands and to "internalize" them or make them his own. The child develops a dislike for playing with feces because his parents dislike him to do this. These internalized prohibitions and self-restraints are considered to be forerunners of the superego. As such, they contain the elements of the future superego, that is, the fear of punishment and the tendency to conform to parental demands. Of course, when parental restrictions are not enforced, the child tends to disregard them.

Superego Development

The actual development of the superego takes place toward the end of the phallic period around age six (see chapter 6). The fear of punishing parents comes to its peak at that point in the Oedipus complex when the little boy, shocked by castration fear, is forced to give up his mother as a love object, and the little girl, under the threat of losing her mother's love, is forced to abandon the father as her love object. The frustrated child of either sex regresses from object relationship to identification by introjection. Parental figures become idealized and seem to be more powerful and more glorious than they may be in reality. In most cases, the father's image plays the greater role in the child's superego. The superego, originally a new element added to and introjected into the ego, gradually becomes a separate mental agency. As such, it is often opposed to the ego, for in contradistinction to the ego, the superego carries a great amount of destructive energy.

The superego, then, becomes the voice of the parents and their moral standards as perceived by the child. It also represents parental wrath and punitive attitudes, and is partly irrational, imposing rigid restrictions not related to present situations.

The ego-ideal, as one of the elements of the superego, carries the child's admiration for his parents. Through the ego-ideal, the superego reflects striving toward perfection and an effort to live up to the expectations of the parents. The child's rebellion and resentment against parental prohibition, however, can hardly find satisfactory discharge, and therefore the superego becomes cathected with some of the destructive energy originally directed against the parents, internalizes it, and directs it against the individual's own ego.

The ego's attitude to the superego is, to a certain extent, a replica of the child's attitude toward his parents. The ego needs love and affection, and its self-esteem depends on the approval of the superego. When the ego lives up to the expectation of the superego, the superego reacts with approval experienced as elation. When the superego disagrees with the ego, however, the aggressive forces stored in the superego turns against the ego, creating guilt feelings.

Identification

The mechanism of identification was described by Freud as follows:

> Identification is known to psychoanalysis as the earliest expression of an emotional tie with another person. It plays a part in the early history of the Oedipus complex. A little boy will exhibit a special interest in his father; he would like to grow like him and be like him, and take his place everywhere. We may say he takes his father as his ideal. . . . At the same time as this identification with his father, or a little later, the

> boy has begun to develop a true object-cathexis towards his mother according to the anaclitic type. He then exhibits, therefore, two psychologically distinct ties: a straightforward sexual object—cathexis towards his mother and a typical identification towards his father. The two subsist side by side for a time without any mutual influence or interference. In consequence of the irresistible advance towards a unification of mental life, they come together at last; and the normal Oedipus complex originates from their confluence. The little boy notices that his father stands in his way with his mother. His identification with his father then takes on a hostile colouring and becomes identical with the wish to replace his father in regard to his mother as well. Identification, in fact, is ambivalent from the very first; it can turn into a wish for someone's removal. It behaves like a derivative of the first *oral* phase of the organization of the libido, in which the object that we long for and prize is assimilated by eating and is in that way annihilated as such. The cannibal, as we know, has remained at this standpoint; he has a devouring affection for his enemies and only devours people of whom he is fond. (1921:121)

It was Freud's conviction that the phallic phase (Oedipal) is comprised of two simultaneous phases (the "negative" phase and the "positive" phase). In the positive phase, identification takes place with the parent of the same sex who is also the child's love-object in the negative phase. In the negative phase there is an identification with the parent of the opposite sex who is the child's love-object in its positive phase. These ambiguous interrelationships could be related to the inherent bisexuality of all children. Freud explained the role of these id identifications in the development of the superego, as follows:

> These identifications [of the positive Oedipus complex] are not what out previous statements would have led us to suspect since they *do not involve the absorption of the abandoned object into the ego*. But this alternative outcome may also occur: it is more readily observed in girls than in boys. Analysis very often shows that a little girl, after she has had to relinquish her father as a love-object, will bring her masculinity into prominence and identify herself with her father, that is, with the object which has been lost, instead of with her mother. This will clearly depend on whether the masculinity in her disposition—whatever that may consist of—is strong enough.
>
> It would appear, therefore, that in both sexes the relative strength of the masculine and feminine sexual dispositions is what determines whether the outcome of the Oedipus situation shall be an identification with father or with the mother. This is one of the ways in which bisexuality takes a hand in the subsequent vicissitudes of the Oedipus complex. The

> other way is even more important. For one gets the impression that the simple Oedipus complex is by no means its commonest form, but rather represents a simplification or schematization which, to be sure, is often enough adequate for practical purposes. Closer study usually discloses the more complete Oedipus complex, which is two-fold, positive and negative, and is due to the bisexuality originally present in children: that is to say, a boy has not merely an ambivalent attitude towards his father and an affectionate object-relation towards his mother, but at the same time he also behaves like a girl and displays an affectionate feminine attitude to his father and a corresponding hostility and jealousy towards his mother. It is this complicating element introduced by bisexuality that makes it so difficult to obtain a clear view of the facts in connection with the earliest object choices and identifications, and still more difficult to describe them intelligibly. It may even be that the ambivalence displayed in the relations to the parents should be attributed entirely to bisexuality and that it is not, as I stated just now, developed out of an identification in consequences of rivalry. . . . As the Oedipus complex dissolves, the four trends of which it consists will group themselves in such a way as to produce a father identification and a mother identification. The father identification will preserve the object-relation to the mother which belonged to the positive complex and will at the same time take the place of the object-relation to the father which belonged to the inverted complex: and the same will be true, *mautatis mutandis*, of the mother-identification. The relative intensity of the two identifications in any individual will reflect the preponderance in him of one or other of the two sexual dispositions. (Freud 1923:59)

In a child whose Oedipus complex is predominately positive, a repression takes place of the incestuous wishes for the parent of the opposite sex and death wishes for the parent of the same sex. At the latency period, the libidinal cathexes toward his mother undergo sublimation, and they express themselves in a nonsexual affection and concern.

Guilt Feeling

The superego as the embodiment of parental prohibitions disapproves of many instinctual desires. Thus, "objective anxiety" is a forerunner of "moral anxiety," for the child's fear of losing his parental love (objective anxiety) becomes transformed into the superego's disapprovals, that is, moral anxiety.

Freud described three conditions under which the repressed instinctual wishes may win access to the conscious:

> The repressed retains its upward urge, its effort to force its way to consciousness. It achieves its aim under three conditions: (1) if the strength of the anti-cathexis is diminished by pathological processes which overtake the other part [of the mind], what we call the ego, or by a different distribution of the cathectic energies in that ego, as happens regularly in the state of sleep; (2) if the instinctual elements attaching to the repressed received a special reinforcement (of which the best example is the processes during puberty); and (3) if at any time in recent experience impressions or experiences occur which resemble the repressed so closely that they are able to awaken it. In the last case the recent experience is reinforced by the latent energy of the repressed, and the repressed comes into operation behind the recent experience and with its help. In one of these three alternatives does what has been hitherto been repressed enter consciousness smoothly and unaltered; it must always put up with distortions which testify to the influence of the resistance (not entirely overcome) arising from the anticathexis, or to the modifying influence of the recent experience or to both (1938:122)

The instinctual desires which were punished by the parents are perceived by the ego as a threat of loss of parental love. After the internalization of parental prohibitions in the superego, a similar process takes place this time under the threat of losing the superego's love. Freud described the instinctual renunciations as follows:

> The chronological sequence would thus be as follows: first, instinct-renunciation due to dread of an aggression by external authority—that is, of course, tantamount to the dread of loss of love, for love is a protection against these punitive aggressions. Then follows the erection of an internal authority, and instinctual renunciation due to dread of it—that is, dread of conscience. In the second case, there is the equivalence of wicked acts and wicked intentions; hence comes the sense of guilt, the need for punishment. (p. 128)

The harshness and severity of guilt in manic-depressive disorder led Freud to believe that the death instinct Thanatos was an intricate and sometimes governing element of the superego. The destructive energy stored in the id flows to the superego as a result of the identification of the ego with the beloved and hated parent and eventual sublimation of libidinal impulses at the conclusion of the Oedipal involvement. The id's libidinal and destructive energy cathexes originally directed toward the parents continue to live on in the superego.

> Every such identification is in the nature of a desexualization or even of a sublimation. It now seems as though when a transformation of this

> kind takes place there occurs at the same time an instinctual defusion. After sublimation, the erotic component no longer has the power to bind the whole of the destructive elements that were previously combined with it, and these are released in the form of inclinations to aggression and destruction. This defusion would be the source of the general character of harshness and cruelty exhibited by the ideal—its dictatorial "Thou Shalt." (Freud 1923:48)

Inner Harmony

As mentioned in chapter 6, in well-adjusted adults there is a "harmonious balance" between the opposing instinctual demands of the id, the self-preservation demands of the ego, and the moralistic demands of the superego. Depressive mood is created by self-directed aggression; it is a result of the ego's being tamed by the superego; manic bliss is caused by a fusion of the ego and superego. But in well-adjusted individuals, the superego plays the role of self-critic, representing the conscience and socially approved norms and standards. To represent the moral code of the individual, the adult superego must outgrow the inital parental prohibitions. The infantile superego represents:

> . . . not merely the personalities of the parents themselves but also racial, national, and family traditions handed on through them as well as the demands of the immediate social milieu which they represent. In the same way, an individual's superego in the course of his development takes over contributions from later successions and substitutes of his parents, such as teachers, admired figures in public life, or high social ideals. (Freud 1938:17)

No longer a replica of infantile images of the parents, the superego of the mature individual becomes more impersonal, related to objective social and ethical standards by which the adult person abides. In well-balanced adults there is little if any conflict between the *moral* standards of the society carried by the superego and the *realistic* considerations of self-protection and survival as represented by the ego.

BIBLIOGRAPHY

Blanck, G. Psychoanalytic Structural Theory. 1977 In B. B. Wolman, ed., *International Encyclopedia of Psychiatry, Psychology, Psychoanalysis and Neurology*. New York: Aesculapius.

Freud, S. Standard Edition. James Strachey, ed. 24 vols. London: Hogarth, 1962; New York: Macmillan, 1962.

——1887–1902. *The Origins of Psychoanalysis: Letters to Wilhelm Fliess, Drafts, and Notes*. New York: Basic Books, 1954.

——1900. *The Interpretation of Dreams*. Standard Edition, vols. 4 and 5.

——1915a. *Instincts and Their Vicissitudes*. Standard Edition, 14:109–140.

——1915b. *The Unconscious*. Standard Edition, 14:159–216.

——1920. *Beyond the Pleasure Principle*. Standard Edition, 18:1–66.

——1921. *Group Psychology and the Analysis of the Ego*. Standard Edition, 18:69–143.

——1923. *The Ego and the Id*. Standard Edition, 19:3–63.

——1926. *Inhibitions, Symptoms, and Anxiety*. Standard Edition, 20:87–174.

——1930. *Civilization and Its Discontents*. Standard Edition, 21:64–145.

——1933. *New Introductory Lectures on Psychoanalysis*. Standard Edition, 22:3–184.

——1938. *An Outline of Psychoanalysis*. New York: Norton, 1949.

——1939. *Moses and Monotheism*. Standard Edition, 23:3–140.

Holt, R. R. 1965. Freud's cognitive style. *American Imago*, 22:163–179.

Sulloway, F. J. 1983. *Freud: Biologist of the Mind*. New York: Basic Books.

[EIGHT]
Social and Cultural Issues

FREUD hypothesized that prehistorical man lived in family groups led by a strong father figure who controlled all members of the family. The most oppressive paternal dictate was that no one was allowed to have sexual relations with the females in the clan. This command forced the brothers to suppress their desires. Finally, the strain of heterosexual abstinence became unbearable. The brothers united, killed the father, and freed themselves from his tyrannic rule and from the ban on sexuality.

After the murder, the original family clan was transformed into a brotherhood of sons who killed the father, cut his body in pieces, and distributed his women amongst themselves. Now, instead of one large family with a tyrannical father, there were several families whose fathers had, to some extent, come to terms with each other; and all of them worshiped in the totem the memory of the archfather. Admiration for the father, combined with hatred for him led to an ambivalent reaction to the father's murder.

Freud made a generalization at this point:

> The totem religion had issued from the sense of guilt of the sons as an attempt to palliate this feeling and to conciliate the injured father through subsequent obedience. All later religions prove to be attempts to solve the same problem, varying only with the stage of culture in which they are attempted and according to the paths which they take. (Freud 1913:10)

Freud defined a group as "a collection of individuals who have introduced the same person in the superego, and on the basis of this common factor have identified themselves with one another in their ego. This naturally holds only for groups who have a leader" (Freud 1933:96). Freud saw the family as the arch-pattern for any group, the group leader as a father substitute, and the group members as united in their admiration for and obedience to the primal father. They all act under the assumption that they are being equally loved by the powerful father. "The indestructible strength of the family as a natural group formation rests upon the fact that this necessary presuppositon of the father's equal love can have a real application in the family" (Freud 1921:140).

Totem and Taboo

In 1913, Freud wrote *Totem and Taboo,* the main purpose of which was to analyze the sociocultural elements of human nature. In this book, Freud related his studies of the individual unconscious to the manifestations of man's collective irrationality.

As a case study, Freud chose the aborigines of Australia. These people do not build houses, they do not cultivate the soil, and they do not keep any domesticated animals other than dogs. They survive on roots and the flesh of wild animals.

Among these people each tribe is divided into small clans, each clan taking the name of its totem, which is either an animal, a plant, or a force of nature. The totem represents the tribal ancestor and protector of the clan. The entire totem system is associated with laws that forbid sexual relations between members of the same totem. Almost all members of the tribe are forbidden to have sexual relations with each other, and these laws are enforced with a strictly administered death penalty. Taboo rules that prohibit incestuous behavior go as far as forbidding young men to speak informally with their mothers, compelling brothers and sisters to avoid each other's company, and compelling married men to avoid their mothers-in-law.

Freud maintained that this incest dread is a subtle infantile trait and is in striking agreement with the psychic life of the neurotic. For Freud, the severity of the punishment imposed indicated how strong the savage's wish for incest must be. He believed that these incestuous desires corresponded to the infantile, incestuous fixations of neurotics in our civilization (Freud 1913).

Freud compared compulsion neurosis to the primitive taboo rules. In both cases, the behavior is unconsciously motivated and seemingly absurd. The compulsive rituals attached to the neurotic and taboo situations are also comparable, as are the mechanisms of contagion and displacement. Compulsion neuroses seems to be the result of an impulse whose gratification was prohibited in early childhood (e.g., an impulse related to touching). The impulse to touch remains with the child, as does the prohibition imposed by the beloved and feared parental figure. By replacing the individual with the group, one can extrapolate a tentative exploration of taboo.

Freud further felt that totemism was similar to the actions of certain children toward animals. Unlike adults, children are usually unable to make a sharp distinction between themselves and animals. Many children develop a considerably more specialized relation: that of a phobia with regard to particular species of animal. Freud found that the etiology of these phobias lies in a displacement of the child's fear of the father. He concluded that totem restrictions and the Oedipus complex have the same underlying motivation, and totemism developed from the Oedipal conflict.

The Future of an Illusion

Human culture, Freud wrote in 1927, includes "all those respects in which man has raised himself above animal conditions." Thus every individual is, in a way, an enemy of culture because of the instinctual renunciations that are expected of him. All cultural institutions, norms, and laws are directed to the defense of the society against the individual's innate destructive and sexual drives. The internalization of the social prohibitions in the superego is a "highly valuable possession for culture."

Prohibitions are a necessary prerequisite for the survival of civilization. If they were removed there would be chaos—one could take whatever one wanted, but many people may want the same. Restraint is necessary. Freud wrote in 1927: "We know already how the individual reacts to injuries that culture and other men inflict on him; he develops a corresponding degree of resistance against the institutions of this culture, of hostility toward it. But how does he defend himself against the supremacy of nature, of fate which threatens him as it threatens all?" Man cannot deal with impersonal forces and he must in some way turn the threat into something familiar to him. Apparently, the only way to influence others is to set up a relationship with them. Thus, "we can have recourse to the same methods against these violent supermen of the beyond that we make use of in our own community; we can try to exorcise them, to appease them, to bribe them and so to rob them of part of their power by thus influencing them" (Freud 1927:8).

Jones explained Freud's ideas as follows: "Psychoanalysis has made us aware of the intimate connection between the father complex and the belief in God, and has taught us that the personal God is psychologically nothing other than a magnified father; it shows us every day how young people can lose their religious faith as soon as the father's authority collapses. We thus recognize the root of religious need as lying in the parental complex" (1961:354).

Freud believed that religion is

> a system of doctrines and pledges that on the one hand explains the riddle of this world to him with an invisible completeness and on the other assures him that a solicitous Providence is watching over him and will make up to him in a future existence for any shortcomings in his life. The whole thing is so patently infantile, so incongruous with reality . . . it is painful to think that the great majority of mortals will never be able to rise above this view of life. . . .
>
> If the achievements of religion in promoting men's happiness, in adopting them to civilization, and in controlling them morally, are no better, then the question arises whether we are right in considering it necessary for mankind and whether we do wisely in basing the demands of our culture upon it. (1927:51)

Religion and Neurosis

Freud explained that every child goes through a neurotic phase. "This is because the child is unable to suppress by rational mental effort so many of these instinctual impulses which cannot later be turned to account, but has to check them by acts of repression, behind which there stands as a rule an anxiety motive." According to the biogenetic principle, humanity goes through a similar neurotic phase. In the ages of mankind's intellectual weakness and ignorance the repression of instinctual desires was of utmost importance. The residue of these repressions, which took place in antiquity, are still with us today. Man's neurosis, like the child's, is not the result of the Oedipus complex. Freud wrote: "Religion would be the universal obsessional neurosis of humanity." The true believer is, therefore, in a high degree protected against the danger of a neurosis, for by participating in the universal neurosis he is spared of experiencing a personal one. The abandoning of religion may take place with the future development and growth of mankind when men will outgrow this phase of infantile dependence and collective neurosis.

Freud has shown in some detail the similarities between religion and the obsessional neurosis in the article "Obsessive Actions and Religious Practices." According to Jones

> Pointing out the same compulsion accompanying the various ritual acts in religious observances (praying, kneeling, etc.) with that accompany-

> ing the private ritual acts of the obsessional neurosis, [Freud] expounded the part played by fear and the sense of guilt if the acts are omitted. They are designed to ward off certain temptations, often unconscious ones, together with the punishments that yielding to them may bring. In the neurosis these are essentially sexual temptations, whereas religious observances are more concerned with aggressive and antisocial ones with conduct in general (1961:353)

There are two types of believers. To the first type belong people who have religious ties of affection. The other type simply obey the rules and regulations because they are intimidated by religion.

Religious beliefs are transmitted from generation to generation. Freud wrote:

> I think it would be a very long time before a child who was not influenced began to trouble himself about God and the things beyond this world. Perhaps his thoughts on these matters would then take the same course as they did with his ancestors; but we do not wait for this development; we introduce him to the doctrines of religion at a time when he is neither interested in them nor capable of grasping their import. . . . We need not be greatly surprised at the feeble mentality of the man who has once brought himself to accept without criticism all the absurdities that religious doctrines repeat to him, and even to overlook the contradictions between them. (1927:83)

The only way that man can control his instincts is through reason and intelligence. But how can one expect man to reach the ideal of intelligence when he is dominated by thought prohibitions. So long as man's early years are influenced by the religous thought prohibitions and by the loyal one derived from it, as well as by the sexual one, we cannot really say what he is actually like. For if we are prepared to renounce our infantile wishes, we can bear it if some of our expectations prove to be illusions. We believe that it is possible for scientific work to discover something about the reality of the world through which we can increase our power and according to which we can regulate our life. Science should guide human life instead of religion and "it would be an illusion to suppose that we could get anywhere else what it cannot give us."

Church and army are good examples of group formation and leadership. The leaders are respectively Christ for the Church and the commander in chief for the army. Christ loves all Christians equally, and the commander in chief loves all his soldiers. The army hierarchy corresponds to the hierarchy of the Church. Members of both organizations have libidinal ties to the leader and to the fellow members. When a tie with the leader is broken, the libidinal ties with members of the group break automatically.

The Discontents of Civilization

Social life is based on renunciation by individuals of some of their libidinal and destructive impulses. This collective restraint is the essence of civilization.

Civilization thwarts the uninhibited outflow of libido. It restrains part of the libido; it allows the use of some parts toward a love object that is really a substitute love object; and it sublimates part of the energy toward art. Aim-inhibited love also leads to the development of friendships and feelings of community belonging.

The love of humanity is an inadequate type of aim-inhibited love. Upon consideration, we realize that if we love blindly, we are not placing much value on our love. It is quite unsatisfactory to love a band of men without good reason. The further one is inhibited away from the prime choice of love objects, the less one gets out of the love relationship. If our limited mental energy were to be put entirely to love, we wouldn't have enough for our own protection. If civilization is to progress, we need to sublimate our sexual energy; and we pay a price for this. On the one hand, love opposes the interests; on the other hand, culture menaces love with grievous restriction.

Sexual restrictions are a necessary source of the energy of noxsexual expression. However, such restriction is not the sole source of impingement on human nature. The bit of truth behind all this—one so largely denied—is that men are not gentle, friendly creatures wishing for love, who simply defend themselves if they are attacked, but that a powerful measure of desire for aggression has to be reckoned as part of their instinctual endowment.

Civilized man has forfeited some of his happiness for a greater chance of security. As time goes on, our civilization will contain less avoidable discomfort, with many deleterious social institutions being eliminated or reduced in unpleasant effect. But there is a certain irreconcilable line of irreducible problems and inherent contradictions: civilization has its price.

Social Restrictions

Society cannot permit full freedom of action for each individual because absolute freedom for one may mean slavery for other individuals.

Curtailment of individual freedom, and inhibition of the instinctual wishes is necessary for the survival of the society. Social norms grew out of this necessity.

> It is quite necessary for a barbarian to be healthy; for a civilized man the task is a hard one. The desire for a powerful and uninhibited ego may seem to us intelligible, but, as it is shown by the times we live in, it is in the profoundest sense antagonistic to civilization. And since the demands of civilization are represented by family education, we must remember to find a place too in the etiology of the neuroses for this biological character of the human species—the prolonged period of its childhood dependence. (Freud 1938:85)

The long childhood of humans and the inability of the human child to face the exigencies of life calls for protective and restrictive actions of the parent. The parental restraint of the child's freedom, and the thwarting of his instinctual wishes must lead to an inner conflict in the child's mind.

At this point, Freud concluded: "We cannot escape the conclusion that neuroses could be avoided . . . if the child's sexual life were allowed free play, as happens among many primitive races." But, on the other hand, this early repression must effect one's readiness for cultural growth, because "the instinctual demands, being forced aside from direct satisfaction, are compelled to take new directions which lead to substitutive satisfaction . . . and may become desexualized." One may conclude "that much of our most highly valued cultural heritage has been acquired at the cost of sexuality and by the restriction of sexual motive forces" (p. 114).

The restriction of the destructive instincts is even more important than the restrictions of sexual behavior. No society could ever survive without instituting definite prohibitions on the use of force. Since inner conflicts and fights could destroy the social organization, taboos have been imposed on the use of force within the boundaries of family and tribe. The thwarted aggressiveness becomes internalized and stored in the superego; from there it may turn against one's own person in acts of self-destructiveness.

Culture means restraint. Social order developed out of restrictions imposed upon two driving forces: sex and destruction. Incest and murder were forbidden. The killed father became totem, and all females inside the tribe became taboo.

Prior to the killing of the father, the sons lived under a permanent castration threat. After killing him, the castration threat became internalized in the form of guilt feelings. Some rituals have to be related to the castration complex. Freud discusses this problem as follows:

> Castration has a place, too, in the Oedipus legend, for the blinding with which Oedipus punished himself after the discovery of his crime is, by the evidence of dreams, a symbolic substitute for castration. The possibility cannot be excluded that a phylogenetic memory trace may contribute to the extraordinarily terrifying effect of the threat—a memory trace from prehistory of the human family, when the jealous father would actually rob his son of his genitals if the latter interfered with him in rivalry for a woman. The primeval custom of circumcision, another symbolic substitute for castration, is only intelligible if it is an expression of subjection to the father's will (compare the puberty rites of primitive people). No investigation has yet been made of the form taken by the events described above among races and in civilizations which do not suppress masturbation among children. (pp. 92–93)

Interpretation of History

The patterns of the development of neurosis are somewhat analogous to the development of religious convictions; and the history of humanity is a history of human desires and passions, sane and insane. Men make history; men and their instincts, emotions, inhibitions, reaction formations, and sublimations determine the course of history. Men have been always lovers and haters; and Thanatos, the god of death and hatred, has never ceased to incite men to wars.

There are two sources of hostility. One is the primary force of threats, of the instinct of death and destruction which may be directed either against self or against the other man. The other is self-love, or narcissism. Narcissism works for the self-assertion of the individual and, in group processes, for the aggressive attitudes of the group toward people who do not belong.

This is why "closely related races keep one another at arm's length. . . . We are no longer astonished that greater differences should lead to an almost insuperable repugnance. . . ." Once cannot be too optimistic about the future of mankind, nor expect a miraculous solution of human conflicts.

Freud was highly critical of abstract interpretations of history, and notably of the "obscure Hegelian philosophy." Man and not "*der absolute Geist,*" men and not dialectics, men and not logical systems make history. Can men harness their instinctual forces and make them work for peace? The future of mankind depends on whether and to what extent the cultural process developed in it will succeed in mastering the arrangements of communal life caused by the human instinct of aggression and self-destruction. In this connection perhaps the phase through which we are passing at this moment deserves special interest. Men have brought their powers of subduing the forces of nature to such a pitch that by using them they could now very easily exterminate one another to the last man. They know this—hence arises a great part of their current unrest, their dejection, their mood of apprehension. And now it may be expected that the other of the two "heavenly forces," the eternal Eros, will put forth his strength so as to maintain himself alongside of his equally immortal adversary (Freud 1930:101).

Freud never dismissed the role of physical environment, nor has he overlooked man's struggle for survival and his breadwinning efforts. However, he thought that a purely economical interpretation of history was too narrow.

> It is probable that the so-called materialistic conceptions of history err in that they underestimate this factor. They brush it aside with the remark that the "ideologies" of mankind are nothing more than results of their economic situation at any given moment or superstructures built upon it. That is the truth, but very probably it is not the whole truth. Mankind never lives completely in the present; the ideologies of the superego perpetuate the past, the traditions of the race and the people, which yield but slowly to the influence of the present and to new developments, and, so long as they work through the superego, play an important part in man's life, quite independently of economic conditions. (Freud 1933:95–96)

Morality

Freud was critical of the idea of a superhuman origin of moral standards:

> The philosopher Kant once declared that nothing proved to him the greatness of God more convincingly than the starry heavens and the moral conscience within us. The stars are unquestionably superb, but where conscience is concerned, God has been guilty of an uneven and careless piece of work, for a great many men have only a limited share of it or scarcely enough to be worth mentioning. (1932:88)

Little children are notoriously amoral; the first source of morality is the parents, who restrain the infant's pleasure-seeking impulses. The fear of punishment and of loss of love and the reward for obedience are the two main sources of children's moral development. To be bad means to do things that annoy parents; and to be good means to do things the parents approve of.

This reasoning is often applied by adults with regard to deity. Many people believe that there is a superior, fatherlike power that rewards people for good behavior and punishes wickedness. Each child is "brought up to know its social duties by means of a system of love-rewards and punishment, and in this way it is taught that its security in life depends on its parents (and subsequently other people) loving it and being able to believe in its love for them" (p. 224).

Moral standards are imposed by the society on the individual and become inner restraints. An individual is "virtually an enemy of culture," and morality must be imposed on him. Moral and cultural restraints stem from without, and the majority of men obey cultural prohibitions "only under the pressure of external force . . . as long as it is an object of fear. This also holds good for those so-called moral cultural demands" (Freud 1927:18).

The second source of moral behavior is love for another person. "Love for oneself knows only one barrier—love for others, love for objects. . . . Love alone acts as the civilizing factor in the sense that it brings a change from egoism to altruism" (Freud 1921:119). Cathexis of libido in others necessarily limits narcissistic love for oneself.

BIBLIOGRAPHY

Freud, S. Standard Edition. James Strachey, ed. 24 vols. London: Hogarth, 1962; New York: Macmillan, 1962.

——1907. *Obsessive Actions and Religious Practices*. Standard Edition. 9:115–128.

——1913. *Totem and Taboo*. Standard Edition. 13:1–162.

——1921. *Group Psychology and the Analysis of the Ego*. Standard Edition. 18:69–143.

——1927. *The Future of an Illusion*. Standard Edition. 21:5–56.

——1930. *Civilization and Its Discontents*. Standard Edition. 21:64–145.

——1938. *An Outline of Psychoanalysis*. New York: Norton, 1949.

Jones, E. 1953–1956. *The Life and Work of Sigmund Freud*. 3 vols. New York: Basic Books.

Wolman, B. B. 1981. *Contemporary Theories and Systems in Psychology*. New York: Plenum.

Wolman, B. B., ed. 1971. *Psychoanalytic Interpretation of History*. New York: Basic Books.

[PART III]
FREUDIAN AND NEO-FREUDIAN THEORIES

[NINE]
Hartmann's Ego Psychology

ONE may distinguish three phases in the evolvement of Freud's concept of the ego. As early as the years 1890–1900 Freud viewed the ego as a mental organization in charge of reality testing, perception, and thinking. *The Interpretation of Dreams* (1900) represents this early theory of the ego.

The years 1900–1922 were probably the most productive years in the development of Freud's theories concerning the instinctual drives, Eros and Thanatos, narcissism, and psychoanalytic technique, but very little that was new was written at that time concerning the ego.

In 1923, Freud published the essay "The Ego and the Id," and he continued for years to come to elaborate the ideas concerning the organization of the ego and its place in the tripartite mental apparatus comprised of the id, ego, and superego. Freud assigned to the ego the key role of a "servant" of the three lords, the id, the superego, and the external reality. The ego was the mediating force trying to keep the balance between the conflicting demands of the id and the superego.

The final revision of Freud's instinct theory (1920) and the Eros-Thanatos dualism permitted the formation of a new ego psychology, but the study of the adaptive functions of the ego continued to be determined by the instinctual frame of reference, namely the perennial struggle between the libidinous life instincts of sex and self-preservation and the death instincts of aggression.

Freud had originally viewed neurotic anxiety as a product of the repression of libido. Since 1923, in accord with his new ideas concerning the functions of the ego, Freud reversed himself and maintained that repression did not *cause anxiety*, but was in fact *produced by anxiety* to ward off the threats coming from within or without. Repression was thus conceived as an escape from danger. Accorning to the revision, anxiety was defined as a signal of distress, but all this made untenable the earlier distinction between the actual neuroses and the psychoneuroses.

The definition of anxiety as a *function of the ego*, rather than as a product of repressed libido, made the early explanation of neurosis based on the vicissitudes of instincts untenable. The neurotic mechanisms had to be reexplained in terms of ego functions concerned with warding off dangers. Freud's solution to the problem was in essence an adaptational one, which apparently contradicted his earlier phylogenetic hypotheses. While retaining the instinctual frame of reference, Freud repudiated the primacy of instincts in molding ego functions.

The problem created thus needed some sort of reconciliation, and several of Freud's disciples endeavored to reassess the structure and functions of the ego (Wolman 1964, 1973).

According to Rapaport, Freud himself laid the foundation for Hartmann's modification of the ego theory. In *The Problem of Anxiety* (1926), Freud ascribed to the ego the ability to make use of the pleasure principle and to turn the passively experienced anxiety into active anticipation. The ego "is ultimately concerned with reality relationships (i.e., adaptation) and therefore curbs instinctual drives when action prompted by them would lead into reality danger" (Rapaport, 1959:10). In 1937, in "Analysis Terminable and Interminable," Freud postulated that the ego has innate elements which are not derived from the id's instinctual forces.

Primary Autonomy

Heinz Hartmann's *Ego Psychology and the Problem of Adaptation*, published in 1939, went beyond Freud's original concept of the ego. Hartmann's ego was no longer a servant; it became the synthesizing power, the overall coordinator of personality.

Hartmann assumed that there are inborn forces in the ego which he called *primary autonomy*. These primary autonomous apparatuses of the ego in maturation constitute the foundation for the ego's relation to external reality. The hereditary core of the ego includes inhibitory forces that delay discharges of energy.

According to Hartmann, the early ego development appears in a new light if one accepts the idea that the ego may be more than a mere by-product of environmental influences on the instinctual drives, and that the ego may have at least a partly independent origin. Hartmann's idea of the autonomous factor in ego development parallels the concept of the id's autonomous drives.

Prior to 1937 psychoanalysts believed that the ego developed out of the growing infant's id drives as a result of contact between the id and external reality. Hartmann stressed that the neonate is endowed at birth with a number of inborn capacities for development. These capacities are above and beyond those that spring from contact between id drives and the environment. These inborn capacities—perception, motility, and memory—have an inherent maturational timetable and do not arise out of conflict. Hartmann called them "ego apparatuses of primary autonomy." Hartmann offered the concept that the infant has only a simple, primary id, viewing the developmental potential at birth as an undifferentiated id-ego matrix (Blanck and Blanck 1974).

In collaboration with Kris and Loewenstein, Hartmann modified the psychoanalytic concept of the id.

> *Functions of the id* center around the basic needs of men and their striving for gratification. These needs are rooted in instinctual drives and their vicissitudes (we do not here deal with these drives themselves and the theory of instincts as developed by Freud). Functions of the id are characterized by the great mobility of cathexes of the instinctual tendencies and their mental representatives, i.e., by the operation of the primary process. Its manifestations are condensation, displacement, and the use of special symbols. (Hartmann et al. 1946:15)

The Origins of the Ego

According to Hartmann, "not all the factors of mental development present at birth can be considered part of the id. . . . I should rather say that both the ego and the id have developed, as products of differentiation, out of the matrix of animal instinct" (1964:119–120). In a paper written in collaboration with Kris and Loewenstein, Hartmann explained that the energies of the aggressive instincts of Thanatos could be neutralized and placed at the disposal of the ego. Aggressive as well as sexual energy may, therefore, be neutralized; and in both cases this process of neutralization takes place through mediation of the ego. Thus, this energy contributes to the development of the ego and makes possible continuing interest in environmental objects regardless of their immediate relation to sexual or aggressive needs (Hartmann et al. 1949).

Hartmann did not assume that the ego is just as inherited as the id; he merely stressed the point that the development of the ego can be traced not only to the impact of reality and of the instinctual drives, but also to a set of factors that are probably genetic and cannot, in any case, be identified with the forces of reality and instinctual drives (Hartmann 1950a).

According to Freud, the id was the only innate part of personality. The ego develops out of the id; the contact with the external reality and the inevitable frustrations and deprivation lead to the formation of a sort of protective shell, called ego. Gradually the ego shifts from the pleasure principle to reality principle and becomes the focal point of the mental system bound to keep in check the instinctual impulses of the id and moralistic demands of the superego.

Hartmann postulated an original undifferentiated id-ego matrix in neonates which gradually develops into id and ego. The development of the ego can be viewed as a process of differentiation that leads to a growing separation between the ego and id, and between the person and the outer reality. It is a gradual development of secondary processes and a transition from pleasure principle of the id to reality principle controlled by the ego. It is also a long road leading from primary narcissism to adult object relations (Hartmann 1952); both the ego and the id developed as by-products of differentiation out of the undifferentiated mass of primitive instincts, and in neonates the id and the ego are one.

Several aspects of the ego can be traced to genetic factors in the id, but the ego development is determined by four factors: inherited characteristics, the instinctual drives, the impact of outer reality, and a complex fabric of factors not identified with either the ego or the id.

Pleasure vs. Reality

While the young child is dominated by the pleasure principle of the id, he gradually learns, whenever faced by displeasure, to adjust the pleasure principle to protect himself. The pleasure principle as such is not a reliable protector of the system, but the primitive ego used the pleasure principle to elicit danger signals in the form of unpleasant feeling, thus fostering the development of the reality principle and self-preservation (Hartmann 1956a). The reality principle includes a postponement of gratification and acceptance of minor displeasure in order to avoid or prevent a major one. The reality principle imposes certain restrictions on the pleasure principle, and also modifies the very conditions for pleasure experiences. This postponement of discharges of emotional energy is the adaptive ability in the neonate, who is endowed at birth with apparatuses of memory, perception, motility, and so on, necessary for coping with the outer world. Thus, Hartmann maintained, "the newborn infant is not wholly a creature of drives; he has inborn apparatuses (perceptual and protective mechanisms) which appropriately perform a part of those functions which, after the differentiation of ego and id, we attribute to the ego" (1939:49).

The Development of the Ego

The development of the ego is a product of both maturation and learning. The innate apparatuses of motility, perception, recall, and so on mature gradually, and at the same time the infants behavior undergoes changes through the process of learning.

This developmental approach to the ego is perhaps the most important contribution to ego psychology, and Hartmann's approach stimulated a host of longitudinal studies.

Hartmann maintained that the differentiation of the ego depends on the infant's ability to separate himself from the world around him. The ability to make this distinction depends on the amount of indulgence to deprivation in meeting the child's instinctual needs, especially food. Hartmann did not agree with Freud's emphasis on deprivation as determinant of separation. According to Hartmann, the child's ability to distinguish himself from his environment depends on the level of cognitive and perceptual maturation. According to Hartmann, in the second half of the first year, the infant gains some control over his own body and begins to anticipate future events. The ability to anticipate the consequences of behavior indicates the transfer from the pleasure principle to the reality principle. Hartmann, Kris, and Loewenstein (1946) maintained that this transition takes place by the differentiation of the ego from the id; the ego becomes gradually an independent variable.

Development of the child's defense mechanisms takes place and the ego assumes the role of protector of life. Some of these defenses, which originally serve the function of adaptation, are rooted in the innate reflex apparatus and may later become mechanisms of defense, which ward off inner anxiety.

The transition from defenses against external threats to the defense mechanisms described by A. Freud is gradual. For instance, the child's identification with his parents is a fundamental factor in the formation of the superego and it serves also as an escape device from the conflicting feelings of love, hate, guilt, and anxiety. According to Freud this identification with the ideal parent is a result of emotional conflict in the phallic stage, but Hartmann maintained that "it rather reflects to concomitant stages of the child's mentation and is probably linked to its original ambivalence" (Hartmann and Kris 1966:33).

Functions of the Ego

Autonomy

The ego's functions develop by learning and maturation. That the ego regulates relations with the environment, that it can organize to find solutions that fit the environmental situation, and the very nature of its psychic system became, therefore, of primary importance for man's self-preservation (Hartman 1948).

The development of the ego, being partly based on the process of maturation, is not entirely traceable to the interaction of drives and environment; indeed, it can become partly independent from the drives in a secondary way. Hartmann terms these factors in ego development primary and secondary autonomy, respectively. The secondary autonomy of functions of the ego has a bearing on the stability of its developmental acquisitions.

The autonomous factors may also come to be involved in the ego's defense against instinctual tendencies, against the outer reality, and against the superego. What developed as a result of defense against an instinctual drive may grow into a more or less independent and more or less structured pattern of behavior. These relatively stable patterns are referred to by Hartmann as *secondary autonomous.*

The relative independence of the ego from the id pressures can be expressed in terms of distance from ego-id conflicts, or distance from the regressive trends exerted by the id determinants. The newly acquired ego functions, the secondarily autonomous, show a high degree of reversibility in the child who uses special devices in his effort to counteract regression.

Hartmann's theory of primary autonomy was supported by research in Albania. Albanian children were tightly swaddled and their motor development was delayed. When they were unbound, their motor development was accelerated and they caught up to normal children. Since motor control is considered one of the ego functions, the development of this function was proven autonomous.

The Conflict-Free Sphere

Not every adaptation to the environment, or every learning and maturation process is a conflict. The "conflict-free sphere" refers to such functions as motor development, perception, attention, object comprehension, thinking, language, and memory.

The autonomous functions of the ego do not always remain in the conflict-free sphere throughout the entire course of development. Sometimes they are influenced by drives; sometimes the autonomous functions become involved in the infant's bodily needs. When the infant anticipates breast or bottle feeding and develops an hallucinatory image of it, a new aspect of ego development has started. The infant's crying is an anticipation of relief rather than an expression of distress; the sensorimotor apparatus of the ego becomes drive-cathected, and the autonomous apparatus of the ego becomes involved with instinctual drives. The primitive connections of the ego apparatus with drive states develop into complex behavorial patterns. The child may start to walk for fun and, at the same time, his walking elicits adult affection. These complex behavioral patterns are formed in the course of development and function autonomously (secondary autonomy).

Adaptation

The main function of the ego is adaptation. The mature ego coordinates the conflict areas and the conflict-free functions, such as motility, perception, and thinking. Human beings are born with a biologically built-in ability for adaptation, called by Hartmann "the state of adaptedness." The "average expectable environment" creates conditions which facilitate the use of the innate potentiality.

Hartmann was concerned primarily with development of functions outside of conflict, such as perception, comprehension, intention, language, thinking, amd motor development. Directly related to this is his concept of the conflict-free ego sphere, which he describes as "an ensemble of functions which at any given time exert their effects outside the region of mental conflicts." The development of the conflict-free sphere largely depends upon the types of defenses used in conflict areas. Some defense mechanisms, such as identification, may become part of primary autonomy.

According to Freud, the interpolation of thought processes enables the ego to delay motor discharge. In the phylogenetic process of evolution, living organisms become more differentiated and less dependent on environmental factors. The trial and error activities became internalized in higher organisms, the frequency of most motor responses is gradually reduced, and intelligence increases. The human species has reached a high level of development and can control the environment through anticipatory intellectual processes. "The proper use of intelligence involves an enormous extension and differentiation of reaction possibilities," Hartmann wrote.

According to Hartmann, intelligence is comprised of three main activities, namely causal thinking, establishing relationships between means and ends, and the ability to accept or manipulate events.

The thinking process leads to rational and purposive behavior, which serves adaptation. The need to perceive reality and to influence it is the fundamental task of thinking. Knowledge must be considered in the biological context of the adaptation problem, for knowing enables the organism to adjust to the environment in a way beneficial for the organism. The knowledge of reality must be subordinated to adaptation to reality, and the knowledge of reality is but the first step in the process of adaptation to reality.

Hartmann viewed intelligence as the organizing and coordinating function of the ego. The inclusion of other mental functions within the ego's methods of thought and action is part of its general anticipatory function which involves man's knowledge of both his relations to his environment and to his inner life. The functioning of intelligence at this level results in better mastery of the environment and better control of one's mind.

At certain points the ego must not be able to cope with the environment. The ego is then forced to exercise its organizing function through increased insight into its own inner world. "Knowledge goes a long way in serving reality but it does not go all the way," Hartmann wrote. To be well adjusted requires flexibility, purposefulness, and adaptability in usual and unusual life situations. "We call a man well adapted if his personality, his ability to enjoy life, and his mental equilibrium are undisturbed" (Hartmann 1958:23).

Intrasystemic Correlations

According to Hartmann, the ego is a complex system and its various inner conflicts are not as significant clinically as those between the ego and the id, or the ego and reality. Psychoanalysts did not think of them as conflicts, at any rate. Hartmann called the intra-ego conflicts "intrasystemic in distinction to the intersystemic, ego-id or ego-superego conflicts."

The intrasystemic correlations and conflicts in the ego have hardly ever been studied. There are several inner conflicts in the ego: the ego has from its start the tendency to oppose the drives, yet one of its main functions is to help to achieve their gratification; then too, insight may border on rationalization, for while it provides objective knowledge, it may convey conventional prejudices of the environment.

The intrasystemic approach becomes essential if such concepts as the dominance of the ego, ego control, or ego-strength are to be clarified. These terms must remain ambiguous until a differential consideration of the ego functions is carefully scrutinized. Consider, for example, the concept of ego-strength. The strength of the ego is commonly judged on the basis of its behavior in typical situations, irrespective of the fact that these situations may be more related to the id, the superego, or outer reality than to the ego proper. According to Hartmann, the autonomous aspect of the ego must be considered. A variety of factors must be taken into account—the strength of the drives, narcissism, tolerance or intolerance against unpleasure, anxiety, guilt feelings, and so on. Hartmann studied the interrelations between the different areas of ego function such as defenses, organization, and autonomy. Whether the defense mechanisms lead to exhaustion of the ego's strength is determined not only by the force of the drive in question, and by the defenses at the ego's frontiers, but also by other factors. A definition of strength must include the autonomous functions of the ego, their interdependence and structural hierarchy and, especially, whether (or how far) they are able to withstand impairment through the processes of defense. This is one of the main elements of how Hartmann conceives of ego-strength. It is probably not only a question of the amount and distribution of ego energy available; doubtless it also has to be correlated with the degree to which the cathexes of these functions are neutralized (Hartmann 1950a).

Change of Function

Hartmann introduced the idea of flexibility in the structural system; he called his idea *change of function.* According to Hartmann, a certain "behavior form which originated in a certain realm of life may, in the course of development, appear in an entirely different realm and role" (1939:26).

In 1950, Hartmann went further. He wrote:

> It seems reasonable to assume that these mechanisms do not originate as defenses in the sense we use the term once the ego as a definable system has evolved. They may originate in other areas, and in some cases these primitive processes may have served different functions, before they are secondarily used for what we specifically call defense in analysis. The problem is to trace the genetic connections between those primordial functions and the defense mechanisms of the ego. Some of these may

> be modeled after some form of instinctual behavior: introjection, to give you but one example, probalby exists as a form of instinct gratification before it is used in the service of defense. We will also think of how the ego can use, for defense, characteristics of the primary process, as in displacement. (1950a:90)

Regression and Neutralization

Before the ego becomes established as an independent organ apparatus, its primitive aims and functions are the result of libidinal and aggressive displacements and symbolizations. During the course of development the cathexis of these aims and functions will become neutralized and attain a certain degree of secondary autonomy. While the instinctual drives still interact with the maturational processes, eventually the ego accumulates its own supply of neutralized energy and can shift them to the points where they are needed.

Neutralization plays a decisive role in the mastery of reality and the individual's utilization of the reality principle. Such secondary autonomous ego functions as thinking, intentional action, and reality testing depend on neutralized energy. Most probably there are also some noninstinctual sources of neutralized energy and some mental energy comes directly from the ego. However, the instinctual drives are indispensable elements in all areas of motivation and adaptation, for without drives there would be no neutralized libido and consequently, no thinking.

The neutralized energy of thinking is responsible for the synthetic as well as the defensive functions, and thought processes perform satisfactory mediation between the id and the ego. High-level cognitive processes originate in the ego and help in the resolution of instinctual desires, thus enabling the individual to function on the reality principle.

Neutralization of energy is to be postulated from the time when the ego evolves into a more or less demarcated structure within the personality. The development of constant object relations presupposes some degree of neutralization. But it is quite possible that the use of this form of energy starts even earlier than at the moment of neutralization just described, and that primordial forms of postponement and inhibition of discharge have previously been fed by a neutralized energy. Some countercathectic energy distributions probably arise in infancy; and there are, probably, transitional states between instinctual and fully neutralized energy. The optimal functioning of the ego depends on the degree of neutralization.

The ego combines the archaic mechanisms of its own and those of the id into an integrated psychic process. The ego does not function only by means of rational and logical processes, but also through primitive feelings and mechanisns. Playful activities, artistic creativity, and fantasies are what Hartmann called a sort of *controlled regression.*

Even mature individuals occasionally engage in playful activities and indulge in daydreaming. Auch an unrealistic behavior is a willfull regression, a process that releases creative energy that may be channeled into fantasy, esthetic enjoyment, and artistic pursuits. The ability to use regressive mechanisms was called by Hartmann "regression in the service of the ego." The psychoanalytic treatment method utilizes controlled regression in the form of transference, in order to resolve past conflicts and strengthen the ego. Temporary ego regressions can stimulate ego growth and facilitate a higher level of functioning. Regression in the service of the ego fosters a productive use of the primitive and archaic mechanisms of both the ego and the id and thus serves the process of adaptation.

This "regression in the service of the ego" can be exemplified by the history of science. Friedrich August Kekulé was unable to establish a formula for the arrangement of carbon atoms in the benzene compound. One night he dreamed of a snake that was swallowing its own tail. Upon awakening, Kekulé realized he had a schematic formula for benzene, having visualized in his dream the structure of the benzene ring.

Theory of Aggression

Hartmann, Kris, and Loewenstein (1949) have introduced a new version of the psychoanalytic theory of aggression and related it, analogously to the libido, to a source, aim, and object. The erotogenic zones serve as sources for sexual stimulation, discharge of energy, and gratification. The destructive instinct uses the body as a tool and as an instrument for the discharge of energy.

> The plasticity of aggression manifests itself in the control of the body, in the control of reality and in the formation of psychic structure. . . .
>
> Libidinal impulses may be aim-inhibited under two conditions: the inhibition may be temporary and may induce an accessory and preparatory stage of impulse completion; or it may substitute for the uninhibited action. In the first case discharge is delayed but under certain conditions mounting pleasure is experienced; in the second case, in which behavior is permanently aim-inhibited, there occurs, in addition to the damming up of libido, substitute-formation or sublimation. (Hartmann, Kris, and Loewenstein 1949:18)

Hartmann, Kris, and Loewenstein conclude their analysis of the aggressive instinct as follows:

> In rounding off our discussion it seems appropriate to enumerate four types of conflict through which the aims of aggression are modified. (1) Aggression and libido may be involved in conflict when the cathexis of both drives is vested in the same object (instinctual conflict). (2) The reaction of the object to attempts at completion of aggressive acts may endanger the individual (conflict with reality). (3) This danger may be anticipated by the ego, which is in part already identified with the object, and the ego might be opposed to the completion of aggressive acts (structural conflict, involving the ego). (4) The conflict may involve moral values (structural conflict, involving the superego). (p. 19)

According to Hartmann, the role of neutralized aggressive energy is at least as important as that of the libido. Aggressiveness may involve the individual in conflicts which may jeopardize his relations with people on whom he depends, and it must be therefore sublimated and eventually integrated into the structure of the ego and superego. The formation of lasting social relations depends on the ability of the individual to bear frustration and sublimate his aggressive impulses. Hartmann and Kris assumed that the capacity to neutralize large quantities of aggressive energy constitutes one of the main determinants of "ego strength" and its capacity for integration (Hartmann and Kris 1949:28).

Hartmann accepted Freud's idea that the ego gets the energy necessary for its needs through sublimated libido. He also believed that the aggressive energy can be changed in a manner analogous to desexualization. Both the libidinal and the destructive energies are necessary for the formation and functioning of the ego, and both of them can be neutralized and thus removed from the instinctual mode of behavior. During the development of the primary autonomous functions, the instinctual functions become neutralized in the process development. Also the secondary autonomy depends on neutralization.

> Neutralization . . . plays a decisive part in the mastery of reality (the prime function of the ego). The formation of constant and independent objects, the institution of the reality principle, with all its aspects, thinking, action, intentionality, all depend on neutralization. (Hartmann 1955:235)

Reality Principle

Hartmann links the reality principle to the way the infant copes with his needs. At the onset, the infant tries to hallucinate away his needs, but when his efforts fail, the infant seeks gratification in reality. Several repetitions of this experience make the infant better acquainted with his physical and social environment, and gradually he learns to master the environment and, whenever possible, to manipulate it for his benefit. The infant applies the reality principle as a means of achieving satisfaction, thus following the pleasure principle. Gradually, through the ego-based reality testing, that reality principle leans toward a set of purposive and cognitive acts. The functions of the ego apparently include postponement and anticipation.

Hartmann's concept of the ego has broadened the scope of psychoanalysis. According to Hartmann, psychoanalysis must become a general psychological system, transcending its earlier scope as a theory of motivation.

> Analysis is gradually and unavoidably, although hesitantly, becoming a general psychology including normal as well as pathological, non-conflictual as well as conflictual behavior . . . and that technique is likely to profit further from this development as it has constantly done since this trend was started by Freud.
>
> Ego psychology has meant a broadening field of view, seeing the connections among facts, the giving of deeper understanding of forms and mechanisms of defense, a more exact consideration of the details of a patient's inner experiences and behavior, and in technique, a tendency toward more concrete and specific interpretation, which includes the infinite variety of individual characteristics and degrees of differentiation which had not been accessible previous to the knowledge of ego functions. (Hartmann 1951:146–148)

Social Aspects

The fabric of social interrelations, the process of division of labor, and the individual's social position determined together the possibilities of adaptation; these qualities also regulate the elaboration of instinctual drives and the ego's development. Social structure determines which behavioral forms will have the greatest adaptive chance. The relation of the individual to his environment is "disrupted" from moment to mement, and must be returned to an equilibrium. Every organism is capable of maintaining or re-establishing its equilibrium and, according to Hartmann, "we can picture the process as an oscillation around the equilibrium" (1939:38).

Hartmann believed that the superego emerges in the Oedipal conflict, but, though genetically related to earlier anal phenomena, it must not be confused with them. The superego is not a part of the system ego, but it does include the ego-ideal. The contents of the superego are differentiated from superego functions; certainly the same applies to their respective cathexes. There may be, moreover, tensions and conflicts between the major systems (id, ego, and superego), as well as conflicts within the superego itself (Hartmann and Loewenstein 1962).

Hartmann discussed the relationship of psychoanalysis to sociology. The psychoanalytic study of social instincts sheds additional light on the basic impulses, as well as on the various types of guilt feelings and their satisfaction by particular social mores and beliefs (Hartmann 1939). The psychoanalytic method may also be of help in the study of social structure and stages of technological development. Sociological studies can benefit from the psychoanalytic theory of conscious and unconscious motivation. The psychoanalytic method can also be applied to the study of national characteristics (Hartmann 1950c). According to Hartmann, "the relationship between the individual and society can be characterized for specific types of people and for specific systems and strata of society not only as to the effect which the system exerts on the individual, but also as to the social functions which the system requires of him" (1950c:29–30). Hartmann believed that the most important future contribution of psychoanalysis to sociology will be that of a general theory of action based on the knowledge of the structural aspects of personality and its motivations (1947).

Synthetic Function

The main function of the ego is the keeping of balance between its inhibitory and adaptive activities. This "synthetic function" of the ego transcends the apparatuses and functions which comprise the ego; the synthetic function makes the ego into the truly coordinating body of personality. The ego gives one the perception of one "being a person," endowed with the ability of experiencing the feeling of unique continuity in time, space, and causal sequence. The ego gradually absorbs functions which have started as defense mechanisms and develops them into secondary autonomy. The development of the ego progresses not only through the mastery of new demand and tasks by creating new apparatuses, but also by taking over on a higher level functions which were originally carried out at a lower level. There is a continuous interaction between the higher and lower levels of functioning. Even the elements of primary autonomy functions are not isolated, and they interact with instinctual drives throughout one's life.

Other functions which Hartmann recognizes as belonging to the ego include the control of discharge motor energy, signals of danger, and the protection of the organism against external and internal threats. Also cravings for material goods and social status are functions of the ego, though they originate in instinctual drives. All these "extra-analytic" areas are included by Hartmann, for adaptation covers both the conflict and the conflict-free issues. The interaction of these two types of processes is a prerogative of normal adaptive behavior. The process of learning to walk upright is a case in point. The acquisition of this skill requires the combination of several factors, such as the innate abilities, maturation of the motor apparatus and, finally, the learning process itself. One must add to these factors the development of the libidinal and identification processes which motivate the child to learn to walk. According to Hartmann, psychoanalysis was predominately a study of instincts and conflict, and must be broadened to include conflict-free areas and adaptation processes.

Mental Health

Sometimes the same processes which represent internal conflict are helpful in attaining reality mastery; they are both pathological and at the same time have a certain adaptational value. Fantasy, which is a denial of reality, is a case in point. Rejection of some aspects of an unpleasant reality is still within the range of normal ego functioning. Fantasy may become pathological when it prevents causal thinking and purposeful and adaptive behavior. Fantasy can, however, have a certain positive and adaptive value and be of some help in problem solving; mental detours can help in coping with reality. Although fantasizing may lead away from adaptation to reality, it may also offer constructive help in adjusting to new situations. Playful fantasy is regressive, but it is a regression in the service of the ego, and it may enable the individual to test (in his imagination) various possibilities without actually experiencing them and taking unnecessary chances.

Hartmann's concept of mental health hinges on adaptation and synthesis. While Freud stressed the importance of instinctual life, the irrational aspects of behavior were emphasized. Hartmann's ego psychology, his new look at defense mechanisms, aggression, and regression in the service of the ego, have introduced the concept of "normal" adjustment. According to Hartmann, mental health depends on four kinds of equilibrium which the organism must maintain, namely between the individual and his environment; between one's various instinctual demands (vital equilibrium); between the three parts of the mental apparatus, the id, ego, and superego (structural equilibrium); and the intrasystemic equilibrium within the ego: an equilibrium between the interdependent parts of the regulators of intrapsychic equilibrium.

Moral Values

According to Hartmann, psychoanalysis as a scientific discipline cannot provide ultimate moral norms, but it may help in clarifying the origins and the meaning of morality. Freud himself lived a highly moral life, and some of his ideas concerning the problems of ethics were formulated by Freud outside the scope of the psychoanalytic theory. Freud studied the roots of moral conduct and expressed certain thoughts concerning the possible relationship between psychoanalytic findings and theories and "Weltanschauugen." While Freud's research dealt with the development of moral attitudes in human beings, Freud himself never fully identified himself with any moral system, nor did he attempt to develop one. Freud traced the "necessity" of moral codes to the fact that human society could not live without them; he called the superego a highly valuable possession for human society; he showed how the development of the superego can lead to a reduction of the external means of coercion used by society; and he noted that "so many people obey only outer pressures" in place of developing their own moral standards. Freud rejected all religious creeds and systems, but he did not reject the moral aspect of the Western civilization. However, Freud stated that "psychoanalysis is not in a position to create a philosophy of life."

Hartmann called the study of values the science of "preferential behavior." Hartmann distinguished three aspects of moral values in relationship to psychoanalysis. First, there is the genesis, the dynamics, and the economics of the patient's moral norms and ideals; second, there is the problem of the confrontation of his moral attitudes with the moral code of his family and of the particular culture he lives in; third, the personal moral valuations of the analyst of the material communicated by the patient.

Psychoanalysis views health as a valid moral criterion. "Health ethics" is related to both the "value irradiation" and "value agglutination." One must become aware of the psychological roots of moral valuation, otherwise one's picture of reality may be distorted and one's judgement of what is healthy and sick may become distorted.

In the development of the superego, sociocultural demands from without, transmitted by parents, are transformed into inner imperatives; usually the character of a command is preserved, and inner tension develops between demand and fulfillment. Not only the prohibitions of the parents but also their love is perpetuated in the relation of the superego with the ego. Every sociocultural system will foster certain types of personal morality. Every system of moral values contains elements related to the demands and the pressures of the superego and to the modifying and more realistic influences of the ego. The individual's moral system is partly conscious, while another is unconscious. Many moral motivations are sheer pretenses, excuses, and rationalizations which conceal the true intentions. Many an individual uses denial or rationalizations against genuine superego commands, and people use repression and reaction formation. The study of the conscious and unconscious processes reveals the motivation and the manner people go about realizing their moral values.

BIBLIOGRAPHY

Blanck, G. and R. Blanck, 1974. *Ego Psychology*. New York: Columbia University Press.

Freud, S. Standard Edition. James Strachey, ed. 24 vols. London: Hogarth, 1962; New York: Macmillan, 1962.

——1900. *The Interpretation of Dreams*. Standard Edition, vols. 4 and 5.

——1915. *Instincts and Their Vicissitudes*. Standard Edition. 14:73–102.

——1920. *Beyond the Pleasure Principle*. Standard Edition. 18:7–64.

——1923. *The Ego and the Id*. Standard Edition. 19:3–63.

——1926. *Inhibitions, Symptoms, and Anxiety*. Standard Edition. 20:87–174.

——1937. *Analysis Terminable and Interminable*. Standard Edition, 22:209–254.

Hartmann, H. 1939. *Ego Psychology and the Problem of Adaptation*. New York: International Universities Press, 1958.

——1947. On rational and irrational behavior. *Psychoanalysis and the Social Sciences*, 1:26–27.

——1948. Comments on the psychoanalytic theory of instinctual drives. *Psychoanalytic Quarterly*, 17:366–388.

——1950a. Comments on the psychoanalytic theory of the ego. *The Psychoanalytic Study of the Child*, 5:74–96. New York: International Universities Press.

——1950b. Psychoanalysis and developmental psychology. *The Psychoanalytic Study of the Child*, 5:7–17. New York: International Universities Press.

——1950c. The application of psychoanalytic concepts to the social sciences. *Psychoanalytic Quarterly*, 19:31–43.

——1952. The mutual influences in the development of the ego and id. *The Psychoanalytic Study of the Child*, 7:9–30. New York: International Universities Press.

——1955. Notes on the theory of sublimation. *The Psychoanalytic Study of the Child*, 10:9–29. New York: International Universities Press.

——1956a. Notes on the reality principle. *The Psychoanalytic Study of the Child*, 11:31–53. New York: International Universities Press.

——1956b. The development of the ego concept in Freud's work. *International Journal of Psycho-Analysis,* 37:425–438.

——1958. Comments on the scientific aspects of psychoanalysis. *The Psychoanalytic Study of the Child,* 13:127–146. New York: International Universities Press.

1960. *Psychoanalysis and Moral Values.* New York: International Universities Press.

——1964. *Essays in Ego Psychology.* New York: International Universities Press.

Hartmann, H. and E. Kris. 1945. The genetic approach in psychoanalysis. *The Psychoanalytic Study of the Child,* 1:11–30. New York: International Universities Press.

Hartmann, H. E. Kris, and R. M. Loewenstein. 1946. Comments on the formation of psychic structure. *The Psychoanalytic Study of the Child,* 2:11–38. New York: International Universities Press.

——1949. Notes on the theory of aggression. *The Psychoanalytic Study of the Child,* 3/4:9–36. New York: International Universities Press.

——1954. Problems of infantile neurosis: a discussion. *The Psychoanalytic Study of the Child,* 9:16–71. New York: International Universities Press.

——1969. *Papers on Psychoanalytic Psychology.* New York: International Universities Press.

Hartmann, H. and R. M. Loewenstein. 1962. Notes on the superego. *Psychoanalytic Study of the Child,* 17:42–81. New York: International Universities Press.

Rapaport, D. 1959. The structure of psychoanalytic theory In S. Koch, Ed. *Psychology: A Study of Science,* 3:55–183. New York: McGraw Hill.

Wolman, B. B. 1964. Evidence in psychoanalytic research. *Journal of the American Psychoanalytic Association,* 12:717–733.

——1973. *Call No Man Normal.* New York: International Universities Press.

[TEN]
Klein's Developmental Theory

MELANIE Klein was a child psychoanalyst. She accepted the main elements of Freud's theory and technique and tried to apply them in the treatment of children. Her modification of the psychoanalytic theory arose out of her clinical experience and mainly applies to child development.

Klein made two original assumptions which do not concur with Freud's theory. The first assumption is that the foundations of the adult personality are laid not in the first five or six years of life as postulated by Freud but mainly in the first year of life. Her second assumption was that the human child is endowed at birth with the ability toward the development of object relations and defense mechanisms. These two highly controversial assumptions, combined with a faithful adherence to almost all psychoanalytic principles as elaborated by Freud, have made Klein and her school into a highly revolutionary and highly orthodox branch of psychoanalysis.

In contradistinction to Freud, Klein assumed that the neonate is endowed with an ego. This assumption brings her somewhat close to Heinz Hartmann. However, while according to Hartmann, the primitive ego is hardly capable of isolating itself from the id, in Klein's theory the neonate's ego is capable of separating itself from objects and relating to them, and also warding off anxiety by an early development of defense mechanisms. According to Klein, the death instinct is the main source of anxiety, and the ego defends itself by using denial, projection, splitting, introjection, and other mechanisms.

The Paranoid-Schizoid Position

The neonate faces at the onset of his life several new situations, mainly the birth trauma, the loss of intrauterine life, and the satisfaction of the breast. A deep anxiety is elicited by the conflict between the two ultimate driving forces, the life and death instincts. The ego protects itself from the anxiety produced by the death instinct by projection and conversion. The ego projects onto mother's breast the part of itself which contains the death instinct. The breast, as the primary object, becomes a persecutor and the source of persecution anxiety. The residues of the death instinct in the ego are converted into an outwardly directed hostility.

Also the libido is projected onto the breast. This projection enables the ego to create the ideal object, which will hopefully satisfy the instinctual needs thus serving the preservation of life. The libido remains at the service of the ego, and a wholesome object relationship is established between the ego and the ideal object. All throughout one's life there is a fusion of the life and death instincts and an incessant interaction of libidinal and aggressive impulses.

The feeling of anxiety arises from both internal and external factors. It is produced internally by the pressures of the death instinct which evokes the fear of annihilation. Externally it is produced by the trauma of birth, and loss of the secure intrauterine state. Klein accepted Freud's idea that the ego projects most of the death instinct outward, and attaches it to the primary objects. A splitting process ensues whereby both the ego and the primary object have been split into good and bad, libidinal and destructive forces.

The infant's earliest experiences of feeding and of his mother's physical presence serve as a start of object-relations. The mother's breast is the object of ambivalent oral-libidinal and oral-destructive attitudes.

The infant's ego attributes all satisfying experiences to the breast viewed as the ideal object, and all the frustrating and painful experiences to the breast viewed as hostile and persecutory object. The infantile fantasies concerning the ideal object and all pleasant experiences are fused with the frustrating experiences and fantasies concerning the hostile object. The ego wishes to incorporate the ideal object, thus securing a never-ending gratification. At the same time, the ego wishes to destroy the persecutory object, the mechanism of denial. Whenever the infant is satiated, a state of balance between the libidinal and aggressive impulses is attained. As soon as the infant feels pangs of hunger or his other needs are frustrated, the aggressive impulses take over. The hungry and greedy infant experiences strong feelings of frustration which strengthen his destructive impulses.

The infant tends to project and introject impulses, and he attributes the libidinal impulse to the "good breast" and the destructive impulses to the "bad breast," and also views himself as the good and the bad breast.

The infant experiences oral-destructive fantasies in which he bites and destroys the breast, and fears that the breast will bite and destroy him. Gradually anal and urethral fantasies prevail and his anxiety is colored by anal and urethral elements.

Defense Mechanisms

In order to ward off the persecutory anxiety, the ego develops several defense mechanisms. At first he uses projection and introjection. When the infant experiences severe anxiety, the split between the ideal object and the bad object is widened, and introjection and projection separate the bad breast from the good breast.

In addition to projection and introjection, splitting and idealization, the ego uses *denial* during the paranoid-schizoid position. The ego not only fantasizes a good and a bad object and idealizes the good one and keeps the good and bad objects wide apart, but often the ego denies the existence of the bad object altogether. This denial of frustration and persecution is tantamount to a denial of psychic reality which is tied to hallucinatory feelings of omnipotence. The denial is used for two purposes, namely the conjuring up of the ideal object and the annihilation of the persecutory object.

Projection enables the infant to rid himself of some of his destructive impulses by deflecting them outwards. The infant attributes his own aggression and hostility to his environment rather than to himself. In this way the infant is able to feel that "They" want to destroy me rather than "I" want to destroy them (Klein 1958: 84).

Also introjection protects the infant against the death instinct. The ego takes in the life-giving food and thereby "binds" the death instinct within (p. 86). Through the processes of projection and introjection the child's first object relations are experienced. As the infant's first object relations are experienced at the mother's breast, the infant learns to love and hate. The mother's breast represents the mother herself. The infant loves the breast because it satisfies all it needs; it removes the child's hunger pains by giving it nourishment and provides for the infant's sexual gratification through sucking.

The oral frustration of the infant leads to oral sadism. Sometimes infants feel that they are being orally frustrated by the mother because someone else is obtaining the gratification they have experienced. An infant who has younger siblings may feel that they have robbed him of the mother's breast. In consequence of oral frustration, the infant may desire to orally incorporate the father's penis. The infant fantasizes that during the sexual intercourse of the parents, the father obtains oral gratification from the mother's breast and the mother obtains oral gratification from the father's penis. The infant may also imagine that both parents obtain oral, anal, and genital satisfaction from each other. This fantasy often leads to a feeling of envy on the part of the infant, who envies the father's penis that is capable of giving the mother babies. Klein maintained that the small child has an unconscious and conscious desire to create babies of its own. As the infant fantasizes that the parents gratify one another, he may feel that this is the reason why his desires are frustrated. When the infant feels that the father has robbed him of the mother's breast, he develops the Oedipus complex and the child may develop the wish for competing with the father for the gratification the breasts provide. The infant may desire to scoop out the mother's breast and enter her body to forcibly obtain the gratification the mother gives to father and withholds from the child.

Most of the aggressive impulses of the young infant are oral but there are also anal and urethral elements. The infant fantasizes expelling excrements out of the self and into the mother. The "good feces" are considered gifts, given to the mother by the loving part of the infant's ego and are believed to be given to the mother viewed as the good object. The bad excrements are parts of the bad ego and are therefore believed to be forced into the mother viewed as a bad object. The mother is not regarded as a separate entity but as a part of the bad self, and the original self-hatred is directed against the mother.

As Hanna Segal has pointed out, "It is the wish and capacity for the restoration of the good object, internal and external, that is the basis of the ego's capacity to maintain love and relationships through conflicts and difficulties" (Segal 1964: 79). A good internalized breast enhances the infant's capacity for integration of the ego. "The good internalized object is thus one of the preconditions for an integrated and stable ego and for good object relations" (Klein 1955: 312). The tendency toward integration is usually present at the earliest phases of the infant's life. The division between the good breast and the bad breast (splitting) must be counteracted in order to develop a healthy ego. Klein believed that the ego possesses an innate capacity to tolerate anxiety; the degree of this capacity depends on inherited strength (Klein 1952: 51) and constitutional factors therefore play a significant role in the development of the ego.

The Superego

The development of the superego starts with the split of the ego in the first year of life (Klein 1955: 399). The superego develops out of introjection of one or both of the parents' images and their demands. The superego "stands over" the rest of the personality. The superego contains a portion of the life and death instincts and parts of the good and bad objects (Klein 1958: 86). Since a part of the death instinct is incorporated in the superego, the death instinct begins to influence the good objects also. The superego entails not only the restraint of hateful and destructive impulses and protection to the good objects from within, but also it includes self-criticism, inhibitions, and persecution resulting from the possession of bad objects. That part of the superego which incorporates the good objects resembles the "good mother," while the other part of the superego, which is ruled by the death instincts, represents the bad mother and her breasts.

Klein assumed that the infant at the paranoid-schizoid position, so called because the leading anxiety is paranoid and the state of the ego and its objects is splitting, which is schizoid, possesses an unintegrated ego, which is capable of experiencing anxiety and of using defense mechanism. The anxiety is caused by the death instinct which produces the fear of annihilation. The ego deflects the death instinct outward to the primary object, the breast, and applies the defense mechanisms of introjection, denial, idealization, splitting, and projection. The infant's introjection of the "good object" reduces the fear, and thus it is of vital importance for his development. If the infant's satisfying experiences dominate over this frustration one, his persecutory fears are kept under reasonable control and he can proceed to the next developmental phase, the depressive position.

The Depressive Position

The above described personality development takes place during the paranoid-schizoid position, which lasts about three to four months of the infant's life. The partial integration of the ego at that stage depends on the amount of gratifying experience which enables the infant to project libidinal impulses on his environment and reintroject them into himself, thus acquiring the feeling of possessing a "good" self. The predominance of gratifying experiences over hostility-producing frustration alleviates the infant's anxiety and leads toward a synthesis of the feelings of love and hostility toward the breast. This synthesis evokes a painful emotion of depressive anxiety and guilt. As the aggression is mitigated by libido, the persecutory anxiety is diminished. However, a new anxiety, related to the fate of the endangered external and internal object leads to stronger identification with it. The infant's ego strives now to make reparation and it inhibits the aggressive impulses believed to be dangerous to the beloved object, Klein wrote.

The process of the splitting at the paranoid-schizoid position protects the integrity of the ego. Splitting meant separating the good experience from the bad one. The process of discrimination and the mechanism of repression originate in splitting. If the splitting during the paranoid-schizoid position is excessive and rigid, the repression will more than likely be of a rigid neurotic nature. If, however, the splitting was less severe, the future repression will not be rigid, and a healthy communication will take place between the conscious and unconscious parts of one's personality.

In the second quarter of the first year of life, a considerable progress in the infant's psychosexual organization takes place. The anal, urethral, and genital elements come to the fore but oral elements still predominate. The infant's ability to express himself and communicate his emotions increases, and his fantasy life becomes more differentiated. The synthetic process reduces the gulf between internal and external objects, and the fusion of libidinal and aggressive impulses underscores the conflict of love and hate, especially towards the infant's mother. The integration of the ego and object proceed simultaneously, the mother becomes a whole object, and the infant's ego becomes a whole ego.

With the progress in integration, the good and bad aspects of the objects come closer together, and the infant realizes that it is one and the same person, that is, himself, who loves and hates one and the same person, his mother. This fusion evokes strong feelings of fear and guilt for he regards his destructive impulses as a threat to the loved object. The feeling of fear and guilt lead to an overall depression experienced by the infant as a threat of losing his loved object. Klein views this fear as an act of mourning.

During the depressive position the infant acquires the recognition of his mother as a total person and he becomes capable of perceiving her as the source of both good and bad. The infant begins to view his mother as an individual who leads a life of her own and who has her own relationships with other people, and not only with the infant. The ambivalent feelings the infant experiences in relation to his mother, and the fear that his aggressive impulses have destroyed her or will destroy her, give rise to a severe anxiety. The infant fears he has lost or may lose the mother he loves and depends upon. The introjection is intensified, for with the depressive position, the infant recognizes his dependence upon his mother and he fears he may lose her. The infant experiences an urgent need to possess the mother and to protect her from his own destructiveness.

The depressive position begins during the oral phase when the cannabalistic love leads to the urge to devour; the omnipotence of oral introjective mechanisms creates a state of anxiety lest these powerful destructive impulses devour both the good internal object and the good external object. The introjected good object forms the core of the infant's ego, thus mourning is experienced because the infant feels he has lost the good object. The outstanding feature of the depressive position is the *feeling of guilt,* because the loss of the love object is believed to be caused by the infant's own destructiveness. The infant feels that he himself destroyed his mother both internally and externally, and now he lost his mainstay of support. The mournful feelings of loss are associated with feelings of guilt and hopelessness in regaining the lost object. Feelings of persecution may also be received, as at the height of the depressive position some regression into the paranoid-schizoid position may take place.

The ego may defend itself against recurring feelings of depression by using defense mechanisms or reparations. The manic defenses are chiefly the same as those used in the paranoid-schizoid position, namely introjection, denial, idealization, splitting, and so on. The manic defenses are less extreme and better organized and they are mainly used to counteract the feelings of depression.

In addition to the defense mechanisms, the ego endeavors to make *reparation* to the injured object. The infant believes that his aggressive fantasies are directed against the love object and he develops guilt feelings combined with the desire to repair, revive, or preserve the love object. This tendency to protect life and to repair damage stems from the libidinal life instinct, and it uses resources in order to reduce depression. The infant's reparation feelings can be described as follows: "My mother is disappearing, she may never return, she is suffering, she is dead. No, she can't be for I can revive her" (Klein 1952: 186).

As the depressive position is worked through the feeling of omnipotence decreases. The infant gradually gains confidence in the objects and his own reparative power. The infant realizes that he can give pleasure to his mother by expressing love. The libidinal expression alleviates the injury wrought by the aggressive impulses, and reparation is made to his love object. Gradually the infant becomes more capable of distinguishing between frustration arising from without and the anguish caused by imaginary internal dangers. The infant's aggression becomes channelled against the external sources of frustration and this more realistic discharge of aggression creates less feelings of guilt; ultimately the infant will learn to channel his aggressive impulses and feelings in an ego-syntonic way.

As the good objects become entrenched within, the infant's relation to people improves, thus giving him a greater feeling of security. His ego is strengthened and enriched, and becomes capable of bringing together and synthesizing the split-off aspects of the objects and of the self. The mechanism of denial, so often applied in the paranoid-schizoid position, is used rarely, and the infant is better prepared to cope with frustrations. During the paranoid-schizoid position the infant used the mother's breast as a scapegoat for all frustrations, but during the depressive position the infant acquires the ability to establish friendly relations with his mother and eventually other people. The infant gains a considerable feeling of security from the new and pleasurable relations. The lessening of the ambivalent feelings facilitate an increased desire for reparation, which in turn, contributes to pleasant feelings of security and the process of mourning is gradually worked through.

The Oedipal Conflict

Although Klein's theory deals mainly with the oral stage of development, her assumption was that the oral frustration may elicit the Oedipus complex as early as the first year of life. Klein assumed that at that early stage, the ego was already capable of projecting and deflecting desires and emotions, experience feelings of guilt and reparation, and develop complex objection relations.

Most often, about the middle of the first year, after reparation has been made, the infant enters into the early stages of the Oedipus complex. He perceives his mother as a whole and separate individual. The infant notices that she is having relations with other people and he discovers that his mother also has a relationship with the father. He projects his own feeling and ascribes to his mother his own oral tendencies. In no time he begins to sense the libidinal link between his mother and father. The infant then projects his own libidinal impulses on both parents, and his desires shift from the mother's breast to the father's penis. He fantasizes his parents in continuous sexual intercourse, giving and receiving gratifications which he is deprived of. The feelings of envy activates the aggressive impulses, and the parents are hated and, in fantasy, also destroyed. This Oedipal envy revives the feeling of depression and paranoid-schizoid attitudes.

One of the infantile fantasies relates to the image of combined parents. The infant is aware of his mother as a whole and separate person but may not be able to differentiate the father from the mother. The mother incorporates all that is good, namely breast, babies, and penises. As the infant's perception of his parents becomes more differentiated and their sexual relationship evokes envy, the infant may regress to denying the father's separate existence and view both parents as if they were one person.

The Oedipal involvement is characterized by frequent fluctuations and conflict between the various libidinal urges.

At the onset of the Oedipal conflicts, both girls and boys desire the mother's breast, but with the gradual working through of the paranoid-schizoid and depressive anxieties, the father's penis becomes the main object of oral cravings. The girls initially desire the penis for oral gratification, but gradually they develop the genital craving and wish to incorporate the penis in the vagina. The boys' desire for the father's penis initially represents a trend toward passive homosexuality. Gradually the urge for father's penis leads to introjection and identification with the father as a person, thus strengthening normal heterosexual development.

The Role of Fantasy

Working at the primitive level of the child's world led Melanie Klein to broaden the concept of unconscious fantasy.

> As the work of psychoanalysis, in particular the analysis of young children, has gone on and our knowledge of early mental life has developed, the relationships which we have come to discern between the earliest mental processes and the later more specialized types of mental functioning commonly called "fantasies" have led many of us to extend the connotation of the term "fantasy" in the sense which is now to be developed. (A tendency to widen the significance of the term is already apparent in many of Freud's own writings, including a discussion of unconscious fantasy.) (Isaacs 1952:79).

Unconscious fantasy springs directly from the instincts and their polarity and from the conflicts between them. Susan Isaacs defined it as "the mental correlate of the instincts" or "the psychic equivalent of the instincts." In the infant's omnipotent world instincts express themselves as the fantasy of their fulfillment. "To the desire to love and eat corresponds the fantasy of an ideal love-, life-, and food-giving breast; to the desire to destroy, equally vivid fantasies of an object shattered, destroyed, and attacking" (Segal 1964). Fantasy in the Kleinian view is primitive, dynamic, and constantly active, coloring external reality and constantly interplaying with it.

> Reality experience interacting with unconscious fantasy gradually alters the character of fantasies, and memory traces of reality experiences are incorporated into fantasy life. I have stressed earlier that the original fantasies are of a crude and primitive nature, directly concerned with the satisfaction of instincts, experienced in a somatic as well as a mental way, and, since our instincts are always active, so a primitive layer of primary fantasies are active in us all. From the core, later fantasies evolve. They become altered by contact with reality, by conflict, by maturational growth. As instincts develop instinct derivatives, so the early primitive fantasies develop later derivatives and they can be displaced, symbolized, and elaborated and can even penetrate into consciousness as daydreams, imagination, etc. (Segal 1964).

This broader concept of fantasy provides a link between the concept of instinct and that of ego mechanism.

> What Freud picturesquely calls here "the language of the oral impulse," he elsewhere calls "the mental expression" of an instinct, i.e., the fantasies which are the psychic representatives of a bodily aim. In this actual example, Freud is showing us the fantasy that is the mental equivalent of an *instinct*. But he is at one and the same time formulating the subjective aspect of the *mechanism* of introjection (or projection). Thus *fantasy is the link between the id impulse and the ego mechanism,* the means by which the one is transmuted into the other. "I want to eat that and therefore I have eaten it" is a fantasy which represents the id impulse in the psychic life; it is at the same time the subjective experiencing of the mechanism or function of the introjection. (Isaacs 1952)

This applies to all mental mechanisms, even when they are specifically used as defenses.

> We are all familiar with fantasying as a defensive function. It is a flight from reality and a defense against frustration. This seems contradictory to the concept of fantasy as an expression of instinct. The contradiction, however, is more apparent than real; since fantasy aims at fulfilling instinctual striving in the absence of reality satisfaction, that function in itself is a defense against reality. But, as mental life becomes more complicated, fantasy is called upon as a defense in various situations of stress. For instance, manic fantasies act as a defense against the underlying depression. The question arises of the relation between the defensive function of fantasy and mechanisms of defense. It is Isaacs' contention that what we call mechanisms of defense is an abstract description from an observer's point of view of what is in fact the functioning of unconscious fantasy. That is, for instance, when we speak of repression, the

patient may be having a detailed fantasy, say, of dams built inside his body holding back floods, floods being the way he may represent in fantasy his instincts. When we speak of denial, we may find a fantasy in which the denied objects are actually annihilated, and so on. The mechanisms of introjection and projection, which long precede repression and exist from the beginning of mental life, are related to fantasies of incorporation and ejection; fantasies which are, to begin with, of a very concrete somatic nature. Clinically, if the analysis is to be an alive experience to the patient, we do not interpret to him mechanisms, we interpret and help him to relive the fantasies contained in the mechanisms. (Quoted after Segal 1967:170ff.)

BIBLIOGRAPHY

Isaacs, S. 1952. The nature and function of fantasy. *International Journal of Psychoanalysis* 29:73–97.

Klein, M. 1921. The development of a child. *Imago*, 7:251–309.

—— 1922. Inhibitions and difficulties in puberty. *Die Neue Erziehung*, 4:69–74.

—— 1923. The role of the school in the libidinal development of the child. *Internationale Zeitschrift für Psychoanalyse*, 9:323–344.

—— 1923. Early analysis. *Imago*, 9:222–259.

—— 1925. A contribution to the psychogenesis of tics. *Internationale Zeitschrift für Psychoanalyse*, 11:323–344.

—— 1927. The psychological principles of early analysis. *International Journal of Psycho-Analysis*, 8:25–37.

—— 1927a. Symposium on child analysis. *International Journal of Psycho-Analysis*, 8:339–69.

—— 1927b. Criminal tendencies in normal children. *British Journal of Medical Psychology*, 7:177–192.

—— 1928. Early stages of the Oedipus conflict. *International Journal of Psycho-Analysis*, 9:167–180.

—— 1929a. Personification in the play of children. *International Journal of Psycho-Analysis*, 10:193–204.

—— 1929b. Infantile anxiety situations reflected in a work of art. *International Journal of Psycho-Analysis*, 10:436–443.

—— 1930a. The importance of symbol-formation in the development of the ego. *International Journal of Psycho-Analysis*, 11:24–39.

—— 1930b. The psychotherapy of the psychoses. *British Journal of Medical Psychology*, 10:242–244.

—— 1931. A contribution to the theory of intellectual inhibition. *International Journal of Psycho-Analysis*, 12:206–218.

—— 1932. *The Psycho-Analysis of Children*. London: Hogarth.

—— 1933. The early development of conscience in the child. In S. Lorand, ed., *Psychoanalysis Today*, pp. 149–162. New York: Covici-Friede.

—— 1934. On criminality. *British Journal of Medical Psychology*, 14:312–315.

—— 1935. A contribution to the psychogenesis of manic-depressive states. *International Journal of Psycho-Analysis*, 16:145–174.

—— 1936. Weaning. In J. Rickman, ed., *On the Bringing Up of Children*, pp. 31–56. London: Kegan Paul.

—— 1937. Love, guilt, and reparation. In M. Klein and J. Riviere, *Love, Hate and Reparation*, pp. 57–119. London: Hogarth.

—— 1940. Mourning and its relation to manic-depressive states. *International Journal of Psycho-Analysis*, 21:125–153.

—— 1945. The Oedipus complex in the light of early anxieties. *International Journal of Psycho-Analysis*, 26:11–33.

—— 1946. Notes on some schizoid mechanisms. *International Journal of Psycho-Analysis*, 27:99–110.

—— 1948. On the theory of anxiety and guilt. *International Journal of Psycho-Analysis*, 29:114–23.

—— 1950. On the criteria for the termination of a psycho-analysis. *International Journal of Psycho-Analysis*, 31:78–80.

—— 1952a. The origins of transference. *International Journal of Psycho-Analysis*, 33:433–438.

—— 1952b. The mutual influences in the development of ego and id. *The Psychoanalytical Study of the Child*, 7:51–53. New York: International Universities Press.

—— 1952c. Some theoretical conclusions regarding the emotional life of the infant. In M. Klein, P. Heimann, S. Isaacs, and J. Riviere, *Developments in Psycho-Analysis*, pp. 198–236. London: Hogarth.

—— 1952d. On observing the behaviour of young infants. In M. Klein, P. Heimann, S. Isaacs, and J. Riviere, *Developments in psycho-analysis*, pp. 237–270. London: Hogarth.

—— 1955a. The psycho-analytic play technique: its history and significance. In *New directions in Psycho-Analysis*, pp. 3–22. London: Tavistock.

—— 1955b. On identification. In *New Directions in Psycho-Analysis*, pp. 309–345. London: Tavistock.

—— 1957. *Envy and Gratitude*. London: Tavistock.

—— 1958. On the development of mental functioning. *International Journal of Psycho-Analysis*, 39:84–90.

—— 1959. Our adult world and its roots in infancy. *Human Relations*, 12:291–363.

—— 1960a. A note on depression in the schizophrenic. *International Journal of Psycho-Analysis*, 41:509–511.

—— 1960b. On mental health. *British Journal of Medical Psychology*, 33:237–241.

—— 1961. *Narrative of a Child Psycho-Analysis*. London: Hogarth.

—— 1963a. Some reflections on *The Oresteia.* In *Our Adult World and Other Essays,* pp. 23–53. London: Heinemann Medical.

—— 1963b. On the sense of loneliness. In *Our Adult World and Other Essays,* pp. 99–116. London: Heinemann Medical.

Segal, H. 1964. *Introduction to the Work of Melanie Klein.* New York: Basic Books, 1964.

—— 1967. Melanie Klein's technique. In B. B. Wolman, ed., *Psychoanalytic Techniques.* New York: Basic Books.

[ELEVEN]
Erikson's Quest for Identity

ERIK Erikson's theories represent a major development of psychoanalytic concepts in relation to cultural and social institutions and moral values. Erikson counterposed his concept of *identity* to Freud's concept of *identification*. According to Erikson, identity "arises from the selective repudiation and mutual assimilation of childhood identifications and their absorptions in a new configuration, which, in turn, is dependent on the process by which a society (often through subsocieties) identifies the young individual, recognizing him as somebody who had to become the way he is." And furthermore, identity is "only one concept within a wider concept of the human life cycle which envisages childhood as a gradual unfolding of the personality through phase-specific psychosocial crises: the epigenetic principle" (Erikson 1968:86).

Despite its sociocultural orientation and deviation from orthodox Freudian concepts, Erikson's system is fundamentally an ego psychology. However, it does not resemble Hartmann's ego psychology. Although Erikson's theory involves the study of the ego, its epigenesis and its self-interpretation, that is, its identity, it is far removed from Freud's system.

Erikson accepted Freud's structural model, and his ego is partly unconscious, partly preconscious, partly conscious. The center of the conscious ego is called the "I" (p. 218). However, while Freud's personality model is transcultural, Erikson's personlity theory and his theory of developmental stages are closely related to the American society and its culture.

The Epigenetic Principle

One of the main tenets of Erikson's theory is the epigenetic principle. The epigenetic principle means that the ego develops out of a general "ground" plan. "Out of this ground plan, the parts arise, each part having its time of special ascendance, until all parts have arisen to form a functioning whole" (Erikson 1959:52).

One can find nothing of that kind in Freud's strictly deterministic philosophy nor in Freud's sober and, perhaps, pessimistic view of humanity. Erikson has never abandoned the fundamental rules of classic Freudian psychoanalysis, but his "epigenesis" and "generativity" brings him close to Adler's style of life and Jung's entelechy.

Erikson's epigenetic principle must be viewed in both phylogenetic and ontogenetic perspectives. Its direct outcome is the principle of generativity which means "the concern in establishing the next generation" (Erikson 1963:267). In a phylogenetic perspective this principle implies the evolution of human behavior from narcissism to caring for other people. Evolution has made man a teaching as well as a learning animal, for dependency and maturity are reciprocal. Mature man needs to be needed, and maturity is guided by the nature of that which must be cared for (Erikson 1968:138). Generativity implies *care* in a historical perspective of a species as well as in the life of an individual. Erikson explained this idea as follows:

> Care is a quality essential for psychosocial evolution, for we are the teaching species. Animals, too, instinctively encourage in their young what is ready for release. . . . Only man, however, can and must extend his solicitude over the long, parallel, and overlapping childhoods of numerous offspring united in households and communities. As he transmits the rudiments of hope, will, purpose, and competence, he imparts meaning to the child's bodily experiences, he conveys a logic much beyond the literal meaning of the words he teaches, and he gradually outlines a particular world image and style of fellowship. All of this is necessary to complete in man the analogy to the basic, ethological situation between parent animal and young animal. All this, and no less, makes us comparable to the ethologist's goose and gosling. Once we have grasped this interlocking of the human life stages, we understand that adult man is so constituted as to need to be needed. (1964:130)

Motivation

Freud's motivational theory postulated innate, instinctual drives and energy systems at their disposal. Erikson's theory of motivation greatly differs from Freud's energetic system. Erikson introduced the concept of *instinctual patterns of behavior*. He abandoned Freud's idea of instinct as being an innate driving force. Instincts, according to Erikson, are some "pre-formed," that is, innate "action patterns." These behavioral patterns can, under certain conditions call on "some ready drive energy for instantaneous, vigorous, and skillful release."

In Erikson's system these "certain conditions" are always determined by social and cultural factors. One may dare to say that in spite of the universal significance of these conditions, Erikson's psychoanalysis is as much American as Sartre's existentialism is French. Erikson's ecological emphasis has its roots in the American situation and the American dilemma. For, as Erikson admitted in his *Autobiographical Notes,* "In this country's history, fate had chosen to highlight the identity question," and further on, "the problems of identity become urgent wherever Americanization spreads" (1970:748).

Actuality vs. Reality

Erikson introduced a significant distinction between Freud's concept of "reality," perceived as the totality of forces and objects in the world, and his own concept of "actuality," perceived as those outer factors which "actuate" the individual, elicit his actions, activate him.

> Mutual activation is the crux of the matter; for human ego strength, while employing all means of testing reality, depends from stage to stage upon a network of mutual influences within which the person actuates others even as he is actuated, and within which the person is "inspired with active properties," even as he so inspires others. This is ego actuality; largely preconscious and unconscious, it must be studied in the individual by psychoanalytic means. Yet actualities are shared, as are realities. Members of the same age group share analogous combinations of capacities and opportunities, and members of different age groups depend on each other for the mutual activation of their complementary ego strengths. Here, then, studies of "outer" conditions and of "inner" states meet in one focus. One can speak of actualities as co-determined by an individual's stage of development, by his personal circumstances, and by historical and political processes—and I will, in fact, speak of all of these. (1962:67)

The Way-of-Life

As mentioned above, Erikson's concept of identity is as much intrapsychic as interactional and social. Identity grows and develops out of the process of mutual actualization and socialization. "The growing child,"

Erikson wrote, "must derive a vitalizing sense of reality from the awareness that his individual way of mastering experience (his ego synthesis) is a successful variant of a group identity and is in accord with its space-time and life plan" (1959:22).

The term "way-of-life" is, actually, Erikson's translation of the German word "*Weltanschauung.*" Erikson interprets this term as being synonymous to "an utopian outlook, a cosmic mood, a doctrinal logic, all shared as self-evident beyond any need for demonstration" (1958:41). Obviously, Erikson believes that such an outlook on life is a necessary prerequisite for the development of identity and, again, he comes quite close to Alfred Adler's theory. Erikson ascribed to the ego the functions Freud partly relegated to the superego. Erikson's concept of the ego is far more inclusive than that of Hartmann's.

According to Erikson, the ego strives toward active mastery of all human desires and experiences. It is an active center of personality. It represents the synthesis of the total personality and it synthesizes the entire experience of a human life.

Developmental Stages

Erikson ascribed particular significance to Freud's developmental phases. The oral phase is characteristic of incorporating, receiving, getting; the anal phase is typical of retention (hold on to) and elimination (let go); the phallic phase represents the masculine mode and the receptive, inclusive feminine mode of behavior, respectively. However, Erikson's concept of bodily zones was quite different from Freud's. Freud's oral zone, e.g. has become the oral-sensory-respiratory.

Small wonder that Erikson postulated definitive developmental stages of an individual in an ontogenetic perspective of the generative man. He linked the individual's development with social norms. An individual must learn to cope with them and be able to overcome the inevitable social crises.

> Man's epigenetic development assures that each of his distinct and protracted childhood stages specializes in one of the major elements (i.e., the numinous, the judicial) which hold together human institutions, each binding together a new set of instinctive patterns and of instinctual energies, of mental and of social capacities so as to assure the continuity of that element, throughout the individual's life and through the sequence of generations. In all epigenetic development, however, a ritual element, once evolved, must be progressively reintegrated on each higher

> level, so that it will become an essential part of all subsequent stages. The numinous element, for example, reappears in judicial ritualizations and in judiciary rituals as the aura which attaches to a personified or abstract image of Justice, or to the concrete persons who as justices are invested with the symbolism and the power of that image. But this also means that neither the numinous nor the judicial elements, although they can dominate a particular stage or a particular institution, can "make up" a ritual all by themselves: always, the whole inventory must be present. (Erikson 1966:602–603)

Freud's developmental stages were, in a way, biologically determined. Every child was supposed to go through the oral, anal, and other stages of development. Freud was aware of the fact that the way the child went through these developmental phases greatly depended on family dynamics and the parent-child interaction. The final outcome, the adult personality, was a product of the interaction between the intrapsychic and social forces.

Erikson went much further in the emphasis on social factors. His developmental stages were determined by the way in which one's ego, at every given stage, was capable of integrating "the timetable of the organism with the structure of social institutions" (Erikson 1963:246).

Erikson distinguished the eight ages of man as follows: basic trust vs. basic mistrust, autonomy vs. shame and doubt, initiative vs. guilt, industry vs. inferiority, identity vs. role confusion, intimacy vs. isolation, generativity vs. stagnation, and ego integrity vs. despair.

Erikson believes that "to understand either childhood or society, we must expand our scope to include the study of the way in which societies lighten the inescapable conflicts of childhood with a promise of some security, identity, and integrity" (p. 277).

At the first stage of life an infant needs continuous care and protection that will enable him to attain a peaceful satisfaction of his basic needs as related to, for instance, the intake of food, bowel movements, and sleep. Motherly care can provide the necessary comfort. "Consistency, continuity, and sameness of experience provide a rudimentary sense of ego identity" (p. 247).

The presence or absence of the feeling of security or, as Erikson calls it, trust, is an important factor in mental health. Parents must convey to the children the basic feeling of trust. The amount of trust does not depend on the quantity of food given to the child, nor on demonstrations of love, but on the quality of maternal care that "combines sensitive care of the baby's individual needs and a firm sense of personal trustworthiness within the trusted framework of their culture's life style" (p. 249).

The second stage (autonomy vs. shame and doubt) is related to muscular maturation of the capacities of holding and letting go (Freud's retention and expulsion). Holding can mean care and protection, but it may also mean restraint and cruelty. Toilet training may lead to a tender care of the infant, or to shaming and doubting him.

> This stage, therefore, becomes decisive for the ratio of love and hate, cooperation and willfulness, freedom of self-expression and its suppression. From a sense of self-control without loss of self-esteem comes a lasting sense of good will and pride; from a sense of loss of self-control and of foreign over-control comes a lasting propensity for doubt and shame. (p. 254)

At the initiative vs. guilt stage the emphasis is on attack and conquest in boys; it is the "phallic-intrusive" mode. In girls it is making oneself attractive and endearing and, in more aggressive modes, it is "catching."

> Infantile sexuality and incest taboo, castration complex and superego all unite here to bring about that specifically human crisis during which the child must turn from an exclusive, pregenital attachment to his parents to the slow process of becoming a parent, a carrier of tradition. Here the most fateful split and transformation in the emotional powerhouse occurs, a split between potential human glory and potential total destruction. For here the child becomes forever divided in himself. (p. 256)

The fourth stage, industry vs. inferiority, corresponds to the child's school age, and the Freudian latency period.

> This is socially a most decisive stage: since industry involves doing things beside and with others, a first sense of division of labor and of differential opportunity, that is, a sense of the technological ethos of a culture, develops at this time. [There is a] danger threatening individual and society where the schoolchild begins to feel that the color of his skin, the background of his parents, or the fashion of his clothes rather than his wish and will to learn will decide his worth as an apprentice, and thus his sense of identity—to which we must now turn. But there is another, more fundamental danger, namely man's restriction of himself and constriction of his horizons to include only his work to which, so the Book says, he has been sentenced after his expulsion from paradise. If he accepts work as his only obligation and "what works" as his only criterion of worthwhileness, he may become the conformist and thoughtless slave of technology and of those who are in a position to exploit it. (pp. 260–261)

The fifth stage, adolescence, is called by Erikson identity vs. confusion. Physiological revolution is but one aspect of the dangers typical of this stage. The ego has to integrate childhood identifications with the vicissitudes of the libido, with endowed aptitudes, and with social roles and opportunities related to the prevailing culture. The adolescent has to develop his sense of identity in his sociosexual role and group identification.

The sixth stage, intimacy vs. isolation, reflects the problems of young adults. Freud was once asked what a normal person should do well. His answer was short: *lieben und arbeiten;* that is, a normal adult is capable of loving and of working. "Satisfactory sex relations," Erikson writes, "make sex less obsessive, over-compensation less necessary, sadistic controls superfluous."

The seventh stage presents the conflict between generativity and stagnation, or productivity and aridity. An individual's mature age may lead to stagnation, or it may open the best chances for productive work and creativity.

The full realization of oneself is presented dramatically by Erikson as ego integrity vs. despair. Lack of ego integration is signified as fear of death. Integrity of the ego is "a postnarcissistic love he partakes" (p. 268).

BIBLIOGRAPHY

Browning, D. S. 1973. *Generative Man: Psychoanalytic Perspectives*. Philadelphia: Westminster.

Coles, R. 1970. *Erik H. Erikson: The Growth of His Work*. Boston: Little, Brown.

Erikson, E. H. 1958. *Young Man Luther: A Study in Psychoanalysis and History*. New York: Norton.

—— 1959. Identity and life cycle: Selected papers. *Psychological Issues*, vol. 1.

—— 1962. Reality and actuality. *Journal of the American Psychoanalytic Association*, 10:451–475.

—— 1963. *Childhood and Society*. Rev. ed. New York: Norton.

—— 1964. *Insight and Responsibility: Lectures on the Ethical Implications of Psychoanalytic Insight*. New York: Norton.

—— 1966. Ontogeny of realization. In R. M. Loewenstein et al., eds., *Psychoanalysis—A General Psychology: Essays in Honor of Heinz Hartmann*. New York: International Universities Press.

—— 1967. Memorandum on youth. *Daedalus*, 96:860–870.

—— 1968. *Identity: Youth and Crisis*. New York: Norton.

—— 1970. Autobiographic notes on the identity crisis. *Daedalus*, 99:740–752.

Evans, R. 1967. *Dialogue with Erik Erikson*. New York: Harper and Row.

[TWELVE]
Mahler's Separation and Individuation Theory

MARGARET Mahler's theory is an outgrowth of close clinical observations of infants and their mothers. She developed her theory of normal and pathological personality formation while she worked with severely disturbed children in a specially designed nursery setting.

Mahler's first principles come quite close to Hartmann's hypotheses concerning the origins of the ego and Freud's theory of primary narcissism. According to Mahler, the newborn child is totally narcissistic, and his mentality as an undifferentiated mass capable of gross reactions to pleasure and pain stimuli.

The ego gradually develops out of that undifferentiated mass. "The ego is molded under the impact of reality on the one hand, and of the instinctual drives on the other" (Mahler 1958:10). However, the infant's intrapsychic processes are greatly influenced by his interaction with the environment and, particularly, his mother. The neonate comes to the world with an unborn apparentness of primary autonomy (cf. Hartmann) and ability to cope with an "average expectable environment, but ultimately, his normal or abnormal development, respectively, depend on the interaction with his mother."

The Autistic Phase

Mahler divided child development in consecutive phases, namely, autistic, in the first few weeks of the infant's life; symbiotic, in the second, third, and sometimes fifth month; and, afterwards, the separation-individuation process begins. Mahler's division is based upon the thoughts of Freud, namely that the child is born in an immature state and that object relations are the most dependable criteria of development.

> In the human young the instinct for self-preservation is atrophied. As a result, the ego has to take over the role of managing the infant's adaptation to reality—a role that the id is unable to fulfill (Freud 1923). The neonate appears to be an almost purely biological organism with instinctual responses to stimuli on a reflex and thalamic level; we can speak only of primitive, unintegrated ego apparatuses. The young infant has only somatic defense mechanisms, consisting of overflow and discharge reactions, at his disposal. The mental apparatus of the neonate and the very young baby is thus not adequate to the task of organizing his inner and outer stimuli in such a way as to insure his survival; it is the psycho-biological rapport between the nursing mother and the baby that complements the infant's undifferentiated ego. Empathy on the part of the mother is, under normal circumstances, the substitute among human beings for those instincts on which the animal is able to rely for its survival. In what is at the start a quasi-closed system, the mother executes vitally important ministrations without which the young child would be unable to survive. During the postnatal period the infant is enveloped in an extra-uterine matrix of the mother's nursing care, a kind of social symbiosis.
>
> From the standpoint of the normal infant, we think that we can correctly describe the very first weeks of extra-uterine life as the stage of "normal autism." In this phase, which extends from birth until the second month of life, the infant makes no discernible distinction between inner and outer stimuli, nor does he seem to recognize any distinction between himself and inanimate surroundings. (Mahler and Furer 1972)

The mental state of an infant in the first few weeks of life is hallucinatory. The infant hallucinates omnipotence, and his entire behavior is directed toward restoration of the "initial" balance or homeostasis. This insistence on sameness and the psychological inability to cope with changes in the environment is normal in the first few weeks of life. Some chilren may never outgrow this phase, and if they relive early infantile autism in later years, they are severely psychotic (see the last section of this chapter).

The Symbiotic Phase

Usually, at the beginning of the second month of life, the infant enters the symbiotic phase. The infant, unable to distinguish between himself and the mother, experiences hallucinatory somatic and mental fusions with his mother. Around the third month of life the primary narcissism somehow seems to recede, and the identification with the mother comes to the forefront. The infant is "absolutely dependent on the symbiotic mother

. . . The infant's need for the mother is absolute," Mahler wrote in 1968.

The mother becomes "symbiotic organizer" of the child's personality. She acts as sort of "auxiliary ego," helping the child to relate himself to the world. She is the first object to be perceived.

> During the time of normal symbiosis, the narcissistically fused object is felt to be "good," i.e., in harmony with with symbiotic self, so that primary identification takes place under a positive valence of love. Later on, after separation, the child may have encountered "bad," frustrating, unpleasurable, even frightening experiences in his interaction with mother and "other," so that the image of the object may have assumed a "negative emotional valence." (Mahler 1971:411)

The term *symbiosis*, borrowed from biology, is defined in that science as the living together of two dissimilar organisms in close association or union, especially where this is advantageous to both; it is to be distinguished from parasitism. Mahler uses it metaphorically: "The essential feature of symbiosis is hallucinatory or delusional, somato-psychic omnipotent fusion with the representation of the mother and, in particular, the delusion of a common boundary of the two actually and physically separate individuals."

Mahler regards both autism and symbiosis as two parts of that phase which Freud designated as primary narcissism. At about the third month of life, with the dim awareness that needs are gratified by an object, symbiosis proper begins.

> Central to Mahler's conclusions from her observation of infants is that optimal symbiotic gratification is essential to development. There can be such extreme communicative mismatching between mother and infant that psychosis ensues. From the side of the infant, mismatching can be the consequence of a defect in the inborn apparatuses. There are infants who are unable to engage in the symbiotic union. (Mahler believes that this observation does away with the commonly held concept of the schizophrenogenic mother). She has also observed the opposite; some infants, especially well endowed, have an unusual capacity to extract from the environment whatever they need for their development. If symbiotic deprivation is severe, the result is symbiotic psychosis or regression to autism. Mahler believes that, actually, childhood psychosis consists of a combination of both symbiotic and autistic pathology. However, the symbiotic psychotic child has some awareness of the symbiotic object, but this is only to the extent of attempting to merge with its "good" aspect and to ward off re-engulfment by its "bad" aspect. In autistic childhood psychosis, capacity to retain some memory traces of good mothering is lost and regression is to objectlessness. With

> an adequate symbiotic experience, however, ego building proceeds. The groundwork for formation of a body image is laid. A rudimentary capacity to mediate between inner and outer perception becomes operative. "The ego is molded under the impact of reality, on the one hand, and of the instinctual drives, on the other." Ego functions are acquired, expecially the important function of delay. This comes about because, with gratification, needs become less imperative. Memory traces of pleasure are linked with perception of the mother's ministrations. This also leads to elaboration of higher levels of object relations. By the second part of the first year, the symbiotic partner has become so specific that she is no longer interchangeable. (Blanck and Blanck 1971:54–55)

According to Mahler, symbiosis with the mother is a necessary phase of development, because at this stage the "immature" infantile organism cannot achieve homeostasis on its own. And, again, what is normal at the first few months of life becomes severely abnormal in fixation and regression at a later phase of life.

The Differentiation Subphase

At the last three months of the first year the infant becomes more active and begins to separate himself from the "fused symbiotic self-plus-object representations" (Mahler 1954). The separation-individuation phase starts with the *differentiation subphase*. The child becomes active, creeps and climbs, tries to stand up straight, and learns to use his entire body.

The maturational growth spurt includes several aspects of the infant's life. The closer the mother is to the infant, the more active he becomes and the more differentiated are his activities. The mother's attention and affectionate attitude facilitate the child's growth and contribute to his ability to separate himself from his mother and grow as an individual.

The infant's attention, which was hitherto *inwardly* directed, gradually gains scope. The infant sleeps less, and in his waking periods he responds to outer stimuli, and especially to his mother. At the age of six or seven months the infant is prone to pat mother's face, to pull her hair, to watch her movements, and to "examine" her nose, mouth, ears, and eyes and to touch her clothes.

> It is during the first suphase of separation-individuation that all normal infants achieve their first tentative steps of breaking away, in a bodily sense, from their hitherto completely passive lap-babyhood—the stage of dual unity with the mother. They stem themselves with arms and legs against the holding mother, as if to have a better look at her as well

as at the surroundings. One was able to see their individually different inclinations and patterns, as well as the general characteristics of the stage of differentiation itself. They all like to venture and stay just a bit away from the enveloping arms of the mother; if they are motorically able to slide down from the mother's lap, they tend to remain or to crawl back as near as possible and play at the mother's feet.

Once the infant has become sufficiently individuated to recognize the mother, visually and tactilely, he then turns, with greater or less wonderment and apprehension (commonly called 'stranger reaction'), to a prolonged visual and tactile exploration and study of the faces of others, from afar or at close range. He appears to be comparing and checking the features—appearance, feel, contours, and texture—of the stranger's face with his mother's face, as well as with whatever inner image he may have of her. He also seems to check back to her face in relation to other interesting new experiences. (Mahler 1972:334–335)

The Practicing Subphase

The second subphase called practicing lasts usually seven to eight months. It starts toward the end of the first year of life. The infant is in an elated mood. He is delighted to be able to move around by crawling or walking. He may become so involved in his activities that he walks away from his mother, although he frequently comes back to the mother for "emotional refueling."

Most of the time the infant acts as if he were omnipotent and totally absorbed in his narcissistic pleasures. He walks upright—and in this new position he can see more things and touch.

During this precious six-to-eight-month period, for the junior toddler (ten-twelve to sixteen-eighteen months) the world is his oyster. Libidinal cathexis shifts substantially into the service of the rapidly growing autonomous ego and its functions, and the child seems to be intoxicated with his own faculties and with the greatness of his world. It is after the child has taken his first upright independent steps (which, by the way, more often than not he takes in a direction away from mother, or even during her absence) that one is able to mark the onset of the practicing period par excellence and of reality testing. Now, there begins a steadily increasing libidinal investment in practicing motor skills and in exploring the expanding environment, both human and inanimate. The chief characteristic of this practicing period is the child's great narcissistic investment in his own functions, his own body, as well as in the objects and objectives of his expanding 'reality.' (Mahler 1972:488–489)

As the child, through the maturation of his locomotor apparatus, begins to venture farther away from the mother's feet, he is often so absorbed in his own activities that for long periods of time he appears to be oblivious to the mother's presence. However, he returns periodically to the mother, seeming to need her physical proximity from time to time.

> The smoothly separating and individuating toddler finds solace for the minimal threats of object loss that are probably entailed in each new stage of progressive development in his rapidly developing ego functions. The child concentrates on practicing the mastery of his own skills and autonomous capacities, continually delighted with the discoveries he is making in his expanding world, quasi-enamored with the world and with his own omnipotence. We might consider the possibility that the elation of this subphase has to do not only with the exercise of the ego apparatuses, but also with the infant's delighted escape from re-engulfment by the still-existing symbiotic pull from the mother.
>
> Just as the infant's peekaboo games seem to turn at this junction from passive to active, to the active losing and regaining of the need-gratifying love object, so too does the toddler's constant running off (until he is swooped up by his mother) turn from passive to active the fear of being re-engulfed by, or fused, with, mother. It turns into an active distancing and reuniting game with her. This behavior reassures the toddler that mother will want to catch him and take him up in her arms. We need not assume that this behavior is intended to serve such functions when it first emerges, but quite clearly it produces these effects and can then be intentionally repeated. (Mahler 1972:491–492)

The child's playful mood may be interrupted by an accident. The infant may fall and hurt himself. The fact that the mother did not prevent it gives rise to an upsurge of separation anxiety and the child runs to mother for help. In contradistinction to Bowlby (1973), Mahler believes that the separation anxiety is a normal and inevitable phenomenon; it is an intrapsychic sensing of a danger signal on the part of the small child during the normal separation-individuation process (Mahler 1968).

According to Mahler:

> At least three interrelated, yet discriminable, developments contribute to and/or, in circular fashion, interact with the child's first steps into awareness of separateness and into individuation. They are: the rapid *body differentiation* from the mother; the establishment of a *specific bond* with her; and the *growth and functioning of the autonomous ego apparatuses in close proximity to the mother.*
>
> It seems that the new pattern of relationship to mother paves the way

for the infant to spill over his interest in the mother on to inanimate objects—at those provided by her—such as toys which she offers, or the bottle with which she parts from him at night. The infant explores these objects visually with his eyes, and their taste, texture and smell with his contact perceptual organs, particularly the mouth and the hands. One or the other of these objects becomes a transitional object. Moreover, whatever the sequence in which these functions develop in the beginning practicing period, the characteristic of this early stage of practicing is that, while there is interest and absorption in these activities, interest in the mother definitely seems to take precedence. We also observed in this early period of practicing that the "would-be fledgling" likes to indulge in his budding relationship with the "other than mother" world. (1972:335)

The Rapprochement Subphase

When the child reaches the age of one year and one half his mental abilities attain the level described by Piaget as concrete, representational intelligence. His ability for verbal communication and manual manipulation of objects rapidly increase. A process of gradual internalization and ego identification begins, and the child becomes increasingly capable of reality testing. However, this development creates new problems, described by Mahler as follows:

Now after mastery of free walking and beginning internalization, the toddler begins to experience, more or less gradually and more or less keenly, the obstacles that lie in the way of what was, at the height of his "practicing," an omnipotent exhilaration, a quite evidently anticipated "conquest of the world." Side by side with the acquisition of primitive skills and perceptual cognitive faculties, there has been an increasingly clear differentiation, a separation, between the intrapsychic representation of the object and the self-representation. At the very height of mastery, toward the end of the practicing period, however, it has already begun to dawn on the junior toddler that the world is *not* his oyster; that he must cope with it more or less "on his own," very often as a relatively helpless, small, and separate individual, unable to command relief or assistance merely by feeling the need for them or giving voice to that need.

During this subphase, some mothers are not able to accept the child's demanding behavior; others cannot tolerate gradual separation—they cannot face the fact that the child is becoming increasingly independent of and separate from them, and is no longer a part of them.

In this third subphase, while individuation proceeds very rapidly and the child exercises it to the limit, he is also becoming more and more aware of his separateness and is beginning to employ all kinds of partly internalized, partly still outwardly directed and acted out coping mechanisms in order to resist separation from the mother. No matter how insistently the toddler tries to coerce the mother, however, she and he no longer function effectively as a dual unit; that is to say, he can no longer get her to participate with him in his still maintained delusion of parental omnipotence. Likewise, at the other pole of the erstwhile dual unity, the mother must recognize a separate individual, her child, in his own autonomous right. Verbal communication has now become more and more necessary; gestural coercion on the part of the toddler, or mutual preverbal empathy between mother and child, will no longer suffice to attain the child's goal of satisfaction, of well-being.

> The junior toddler gradually realizes that his love objects (his parents) are separate individuals with their own individual interests. He must gradually and painfully give up his delusion of his own grandeur, often through dramatic fights with mother, less so it seemed to us, with father. This is a crossroad that we have termed the "rapprochement crisis." (Mahler 1972:494–495)

Apparently, there is a price to be paid for a broader conception of the world and oneself. At this third subphase of the separation-individuation process, the child begins to perceive himself in a more realistic vein. He may become aware that he is rather weak and lonely. While his real power has increased, the former hallucinatory omnipotence gradually disappears. The active and outgoing child feels an increasing need for maternal support.

Whenever he is hurt or upset, he turns to his mother to make sure that she is with him, ready to protect and help. While he becomes aware of his growing independence, he willfully solicits her attention by standing behind her chair or demanding she hold him or, at least, that she constantly pay attention to him.

If a mother is "emotionally and quietly available" and offers the child adequate supply of object libido, she facilitates his normal development.

Symptom Formation

The less attention the mother gives to the child at this phase the more insecure he will become and the more desperate his demands. He may lose interest in his toys and follow his mother around the house or he may become overtly hostile and destructive.

The main symptoms created by maternal rejection at the rapprochement phase are *shadowing, darting away*, and *separation anxiety*. According to Mahler, the first two characterstic patterns of behavior, the shadowing of the mother and the darting away from her expecting that she will chase the child and take him in her arms are indicative of "the toddler's wish for reunion with the love object and . . . fear of re-engulfment" (Mahler 1971:411). The toddler fears to lose maternal love and, at the same time, his autonomy. The negativism and aggressiveness of Freud's *anal phase* will coincide with Mahler's "darting away" at the *rapprochement phase*.

> The less gradually the intrapsychic separation-individuation process takes place, and the less the modulating, negotiating function of the ego gains ascendancy, the greater the extent to which the object remains an unassimilated foreign body, a "bad" introject in the intrapsychic emotional economy. In the effort to eject this "bad" introject, derivatives of the aggressive drive come into play and there seems to develop an increased proclivity to identify with, or to confuse, the self-representation with the "bad" introject. If this situation prevails during the rapprochement subphase, then aggression may be unleashed in such a way as to inundate or sweep away the "good" object, and with it the "good" self-representation. This would be indicated by early, severe temper tantrums, for example, in children in whom the too sudden and painful realization of their helplessness results in the too abrupt deflation of their previous sense of their own and shared magic omnipotence (in Edith Jacobson's sense, 1964).
>
> I observed many of our normal children recoil, or show signs that had to be interpreted as a kind of eroticized fear, on being cornered by an adult who wanted to seek, often playfully, bodily contact with the child. This seemed to be felt as overwhelming by the toddler because of the adult's sheer bodily size and strength.
>
> These behaviors remind us of the fear of re-engulfment by the by-then already somehow contaminated, dangerous "mother of separation" in whose omnipotence the child still believes, but who does not seem to let him share in her omnipotence anymore.
>
> In incipient infantile neurosis, conflict is indicated by coercive behaviors directed toward the mother, designed to force her to function as the child's omnipotent extension. This alternates with signs of desperate clinging. In other words, in those children with less than optimal development, the ambivalence conflict is discernible during the rapprochement subphase in rapidly alternating clinging and increased negativistic behaviors. This may be in some cases a reflection of the fact

> that the child has split the object world, more permanently than is optimal, into "good" and "bad." By means of this splitting, the "good" object is defended against the derivatives of the aggressive drive. (Mahler 1971:412–413)

Toward Object Constancy

The fourth subphase of separation and individuation starts at the end of the second year and lasts all third year of life. At this period "object constancy is attained," and the child begins to develop two separate concepts of himself and his mother.

At this period, the child develops more independence from his mother. He may leave the room where the mother is and play by himself. For a few hours a day he may accept the nursery teacher as sort of a mother surrogate. He begins to take interest in children and adults besides his mother.

The child's attitude is often negativistic however, there is no longer need for splitting and hostile behavior.

Gradually, the child's personality and behavior become more integrated. His affectomotor and sensorimotor patterns become more harmonious. The "good" and "bad" self and object images blend, the aggressive impulses neutralized, and the parental images become internalized, and the id, ego, and superego emerge in a wholesome interaction. The child overcomes separation fears and grows toward eventually becoming one individual.

Child Psychosis

The infant's failure to use his mother as an organizer of his maturation may cause grave harm. The maturational failure leads to child psychosis and its two syndromes, autistic and symbiotic.

> In the anamnesis of the earliest behavior of the autistic psychotic child, one learns that he showed no anticipatory posture at nursing, nor was there any reaching-out gesture or specific smiling response. Many such children exhibit what amounts to exclusive attachment to an inanimate object such as a toy or a piece of plastic; or else they display a primary and exclusive preoccupation with their own body, via rocking, head-rolling, or violent banging of the head, injurious scratching of their skin, a continuous fascination with their own fingers, etc. Unable to utilize the auxiliary executive ego functions of the symbiotic partner, the mother,

> the child has had to develop substitutive modalities of orientation in order to be able to cope with stimuli and to survive. He has thus created and enclosed himself in a quasishell of a small restricted world of his own which offers him the fulfillment of his imperative "demand for sameness" (Kanner 1944). He is therefore either completely mute or his language serves only as a signal directed toward the seemingly de-animated and de-differentiated, mechanical world (which includes the mother).
>
> Autism constitutes an attempt to achieve de-differentiation and de-animation; it serves to counteract the multitudinousness and the complexities of external stimuli and inner excitations which threaten the autistic child's existence as an individual entity. Without the partnership of his mother, such as the normal child depends upon, he lacks the capacity to deal with the complexity of stimuli (which engenders in him an overwhelming anxiety and annihilation panic). Any attempt to impinge on the negative hallucinatory barrier that he is intent on maintaining results in a panic-rage attack. As part of the de-differentiation process, which is in the service of the maintenance of life, there is an absence of distinction between his own body and the world around him; it would seem that there is also a relatively low cathexis of his body surface as compared with his investment of his visceral organs. This results, on the one hand, in insensitivity to pain, and on the other, given the typical preponderance of the aggressive drive, in psychosis, a turning against the own body which culminates not infrequently in autoaggressive and self-mutilating activity. (Mahler 1958:78–81)

On the other end of the spectrum of child psychosis, there is the syndrome of symbiotic psychosis (Mahler 1968). In cases of this kind the symbiotic phase of development has been attained, albeit in a grossly distorted fashion. But disturbances in maturation and/or development have seriously interfered with the progress of the subsequent separation-individuation process, and the result has been fixation at or regression to a distorted symbiotic phase. Clinically, we do not observe a turning away from the human reality in infancy, such as we find in the autistic syndrome: a more normal development during the first two years in terms of emotional responsiveness is reported in the anamnestic data. Yet close scrutiny of the developmental history often reveals unevenness of growth, together with a striking vulnerability of the budding ego to minor frustrations. For example, such children will give up locomotion for months because they fell down when they were making their first attempts at walking.

> The psychosis itself is often ushered in by acute panic evoked by such routine separation experiences as enrollment in nursery school, birth of

a sibling, and so on. Our belief, however, is that basic to the disturbance is the child's reaction to the inherent maturation pressures toward intrapsychic separation from the mother. That inherent pressure is manifested behaviorally, by the inevitable onset of autonomous locomotion. Thereafter, the clinical picture may be dominated by agitated catatonic-like temper tantrums and panic-stricken behavior, followed by bizarrely distorted efforts toward reality testing and hallucinatory modes in the direction of restitution. Underlying all the different symptoms are the child's attempts to restore and perpetuate the delusional omnipotent mother-infant fusion that existed during the symbiotic period. In the fully developed symbiotic syndrome, echolalia and echopraxia make their appearance—either derived from the inanimate world (e.g., television commercials) or, at times, from the animate world, chiefly the mother. There may be attachments to inanimate objects which we have named "the psychotic fetish" (Furer 1964). By contrast with the autistic syndrome, there is a rather complex ambivalent or "ambitendent" relationship on the child's part to these substitutes for the symbiotic partner. In many cases of the primarily symbiotic syndrome, because of his continuing panic-stricken state, the child is compelled to take recourse to a secondary retreat into a quasi-stabilizing autism. However, aggressive destructive behavior, directed toward both inanimate and animate objects, may also make its appearance. (Mahler 1972:215–216)

BIBLIOGRAPHY

Bowlby, J. 1973. *Attachment and Loss.* 2 vols. New York: Basic Books.

Furer, M. 1964. The development of a preschool symbiotic boy. *The Psychoanalytic Study of the Child.* New York: International Universities Press.

Jacobson, E. 1964. *The Self and the Object World.* New York: International Universities Press.

Mahler, M. S. 1951. On child psychosis and schizophrenia: Autistic and symbiotic infantile psychosis. *The Psychoanalytic Study of the Child,* 7:286–305. New York: International Universities Press.

—— 1958. Autism and symbiosis, two exreme disturbances of identity. *International Journal of Psycho-Analysis,* 39:77–83.

—— 1961. On sadness and grief in infancy and childhood: loss and restoration of the symbiotic love object. *The Psychoanalytic Study of the Child,* 16:332–351. New York: International Universities Press.

—— 1963. Thoughts about development and individuation. *The Psychoanalytic Study of the Child.* 18:307–324. New York: International Universities Press.

—— 1965. On the significance of the normal separation-individuation phase. In M. Schur, ed., *Drives, Affects, and Behavior,* 2:161–168. New York: International Universities Press.

—— 1966. Notes on the development of basic moods: the depressive affect in psychoanalysis. In R. M. Loewenstein, L. M. Newman, M. Schur, and A. J. Solnit, eds., *Psychoanalysis—A General Psychology,* pp. 152–168. New York: International Universities Press.

—— 1968. *On Human Symbiosis and the Vicissitudes of Individuation.* New York: International Universities Press.

—— 1971. A study of the separation-individuation process: And its possible application to borderline phenomena in the psychoanalytic situation. *The Psychoanalytic Study of the Child.* New Haven: Yale University Press. 26:403–424.

—— 1972. On the first three subphases of the separation-individuation process. *International Journal of Psycho-Analysis.* 53:333–338.

Mahler, M. S., and P. Elkisch. 1953. Some observations on disturbances

of the ego in a case of infantile psychosis. *The Psychoanalytic Study of the Child,* New York: International Universities Press. 8:252–261.
Mahler, M. S. and M. Furer. 1963. Certain aspects of the separation-individuation phase. *The Psychoanalytic Quarterly,* 32:1–14.
—— 1972. Child psychosis: a theoretical statement and its implications. *Journal of Autism and Childhood Schizophrenia,* 2/3:213–218.
Mahler, M. S. and J. B. McDevitt. 1968. Observations on adaptation and defense in *statu nascendi. Psychoanalytic Quarterly,* 37:1–21.

[THIRTEEN]
Wolman's Interactional Theory

FREUD'S work could be divided into three major areas, namely empirical observation, theoretical contructs, and application of the two toward the treatment of mental disorders. Keeping within the framework of the present volume, the description of my writings will be limited to my contribution to the second area only.

I adhere to Freud's ideas of unconscious motivation, and the structural and topographic theories, but introduce certain modifications in the theory of drives, female psychology, social relations, developmental stages, and psychopathology.

Eros and Ares

Where there is life, there is energy which can be used either to support or to destroy it. The energy of a living organism can be directed toward protection or destruction of oneself or others. Life represents amounts of energy; the life instincts of love and hostility indicate the aim of this energy and the direction in which it is channeled. This two-dimensional relationship can be presented graphically on Cartesian coordinates, with the vertical dimension representing the amount of energy, and the horizontal line indicating the use of energy; the plus direction means the protection of life, the minus direction the destruction of life. The distinction is made between the amount of energy one has (i.e., how much life and vitality one possesses) and the *way* in which this energy is used (i.e., for or against survival).

The intake of food is done for the protection of one's own life. This is, however, frequently accomplished at the expense of someone else's life. Every living organism needs food for survival and every living organism seeks food. To get food is a matter of life for the eater, and a matter of death for the one to be eaten up. Wolves and sheep are the proverbial poles of this life and death axis, but sheep eat, too. Sheep, being herbivorous, destroy plants by eating them. Wolves, being carnivorous, destroy sheep by eating them. The wolf who eats the sheep does not serve Thanatos; he eats for his own survival. The sheep who is eaten up has not done anything to promote its own death, but fell prey in the fight for survival. Apparently, the greatest single destructive action, eating, does not conform to Freud's theory of Thanatos.

Freud assumed that self-destructiveness precedes object-directed destructiveness. When outward aggression is unable to find adequate gratification in the external world, it may "turn back and increase the amount of self-destructiveness within." It would seem, therefore, "as though it is necessary for us to destroy some other thing or person in order not to destroy ourselves, in order to guard against the impulsion to self-destruction" (1933:105).

Freud's theory of Thanatos implies that suicidal tendencies occur earlier than the genocidal tendencies, in both the phylogenetic and ontogenetic sense. Were this true, animals would have eaten themselves up before they ate one another, and babies would have hurt themselves before they began to suck the mother's breast.

There are cases of self-directed hostility. According to Freud, the self-directed hostility is seated in the superego, and the superego is the last of the three systems of the mental apparatus to develop. The superego is invested with parental hostility toward the child and the child's hostility toward his parents, and both types of hostility are object-directed. They may become self-directed by the introjection of parental images. Thus, according to Wolman's interactional theory, object-directed hostility is primary and self-directed hostility is secondary.

Self-Preservation

This tendency of living matter to restore equilibrium has been elevated to a most general biological law by Cannon, Goldstein, Pavlov, and Freud. The only qualification is to be made at this point is that the homeostatic principle, whether in Pavlov's, Goldstein's, or Freud's version, is a universal law of living matter. Thus, the main thesis of the interactional theory reads:

> All living matter is endowed with biochemical energy derived from the universal energy that, in turn, as explained by Einstein, is a derivative of matter. At a certain evolutionary level this biochemical energy

> is transformed into mental energy. This mental energy serves survival. The apparatus of discharge, call it drive, instinct, or instinctual force, reflects the most universal urge to stay alive. It is the "Lust for Life," the wonderful craving of all living matter to live. (Wolman 1973:37)

Self-preservation is neither a myth nor a hypothesis; it is the most general and best known empirical fact pertaining to living matter. The various functions that serve self-preservation may be divided into intake of oxygen, food, and water, flight, fight, and so on. One may call oxygen, food, and water "needs," defining "need" as a condition for survival. Deprivation of one of these or other biological needs, defined as conditions for survival, motivates the organism to act in a way that leads toward the satisfaction of these needs. The self-preserving actions are, as a rule, typical for each biological species and are determined by heredity, anatomy, and physiology of each organism, by its age and physical fitness, by physical and social environment, and also by intelligence, learning, and other factors.

The interactional theory assumes that the observable behavior of all living organisms is *conservative* in the sense of the conservation or preservation of life. This overall tendency, described by Charles Darwin as the "fight for survival," can be conceptualized in terms of an overall instinct or drive or complex of unconditioned reflexes, corresponding to Pavlov's instinct of life. Pavlov wrote: "All life is nothing other than the realization of one purpose, viz., the preservation of life itself, the tireless labor of which may be called the general *instinct of life* (1928:277). And further on: "Life is beautiful and strong only to him who during his whole existence strives toward the always desirable but ever inaccessible goal. . . . All life, all its improvement and progress, all its culture are effected through the reflex of purpose, are realized only by those who strive to put into life a purpose" (1928:279).

The functions of such a general instinctual drive for life can be presented in the following manner. Each organism has at a certain time a certain quantity of energy. The fact that matter, in accordance with Einstein's formula $E = mc^2$, is a peculiar accumulation of energy does not contradict this line of reasoning. The energy is discharged, in accordance with the principle of equilibrium, whenever the organism is exposed to internal or external stimulation.

Lust for Life

The force which activates energies in the direction of survival should be named "lust for life." At a certain evolutionary level part of this energy has become invested in procreation or the preservation of the life of the species as if life of one individual continued through this offspring. From that time on the "lust for life" drive split into Ares and Eros, the War and Love drives respectively.

As long as an organism is alive, its energies can be used in two directions, either toward the promotion of life or toward its destruction. The instinctual force, "lust for life," divides into two arms; the one that serves promotion of life is *Eros*, or love. Love can be directed toward oneself or toward others. The other arm serves destruction and may be called *Ares*. Ares, too, can be directed toward oneself or toward others. Life and death deal with *quantities* of energy. The drives of love and hostility indicate the *direction* in which the mental energy is used by Eros and Ares.

Apparently, Ares is not only a more primitive and phylogenetically older drive than Eros, but it is probably more powerful than Eros. Pavlov's dogs did not copulate when their skin was burned, but they salivated even when he burned their skin (Pavlov 1928:228). It seems that the cortical food center is stronger than the skin center, and the sex center is weaker than the skin center. When Pavlov tried to crush the dog's bones, even salivation stopped. Hungry dogs can bear minor wounds, but bone breaking means death, and in face of death all energy is mobilized for self-defense.

Hungry, thirsty, sick, and wounded organisms act in an aggressive and destructive manner. It seems that whenever the supply of libido is used up, the organism works on destrudo. Perhaps the libido is sort of a "higher," and the destrudo a "lower" brand of the same fuel. In danger and anger men act with what seems to be added energy, the latent energy of destrudo.

Eros and Ares are the two basic releasers of mental energy. Ares, like Eros, has an impetus, source, object, and aim. The impetus is the amount of destructive energy (destrudo) that is discharged. Its source is a threat to one's own life. The aim of Ares is the complete or partial destruction of enemies. The object can be oneself or any other organism.

Ares, the instinct of hostility, fits well into the definition of instinct given by Fenichel: "The instinct attempts to remove the somatic changes at the source of the instinct" (1945:60). The somatic changes in Ares are an accelerated heartbeat, perspiration, trembling, contraction of muscles, baring of teeth, growling and so on. The threat of annihilation is the main cause of the somatic changes. The hostile action, which is a discharge of destrudo aiming at the destruction of the threatening object, restores the inner balance (analogous with the actions of Eros). The threat may be related to an inner stimulus of hunger or an outer stimulus that jeopardizes one's life, prevents satisfaction of hunger, prevents escape, or any other combination of hostile stimuli.

Pavlov (1928:255 ff.) believed that hostile behavior is a *guarding reaction* against real or threatened injury. Experimental studies on frustration (Dollard et al. 1939; Lawson 1965) stressed the fact that frustrated individuals tend to become hostile.

Love, Sex, and Power

This shift of emphasis from Eros, libido, and sexuality in the direction of survival and adjustment to life follows, to a certain extent, the ideas of Kardiner, Hartmann, and Erikson, but Wolman's interactional theory reinterprets several aspects of psychosexual development and sexual behavior in the light of the fundamental *lust for life* drive. In the phylogenetic development of nature, the fight for survival was the earliest and most fundamental driving force, while sexual procreation appeared much later in the process of evolution. With very rare exceptions, *the urge to live is more powerful than the urge to love,* and sexual behavior is greatly influenced by the more basic needs related to survival.

Marriage and family are institutions primarily geared to sexual relations and child rearing, yet the economic factors are of utmost importance in marital contracts, child care, and survival or collapse of the family organization. Often women married for the sake of financial security, and men expected fat dowries, and the children's well-being depended on parental power, that is, their ability to satisfy the children's needs. Thus, the concept of power introduced by me (1966a, 1981) has modified Freud's theory of sex.

The assumption that fight for survival is the main force in most if not all higher organisms led Darwin to the postulation of two innate drives, namely, the instinct of survival of the species and the instinct of survival of the individual. Several scientists, among them Sigmund Freud, followed in Darwin's footsteps. Freud originally (1887–1902) postulated ego instincts related to the survival of the individual and libido instincts related to the survival of the species. Since all higher species procreate in a sexual manner, sexuality was believed to serve the survival of the species.

In the "Three Essays on Sexuality" (1905) Freud clearly distinguished between procreation and human sexual drive. The human sexual drive was freed by Freud from the procreative bind provided by Judaism and Christianity; Freud established the sexual drive as an independent and all-important motivating factor related to the search for pleasure and the libido has become the single greatest driving force and the determing factor in psychological growth and development from birth to maturity.

Freud's theory of psychosexual development included most aspects of human behavior, and the developmental stages of the libido were presented as the key factor in personality structure. Terms such as the "oral character," "polymorphous pervert," and "object love" bear witness to the all-important role ascribed by Freud to sexuality in personality development and human relations (Freud 1898, 1931; Wolman 1981a; Wolman and Money 1980).

I maintain that Freud's ideas concerning sexuality were based on observations of a *particular* phenomenon in a *particular* cultural-historical setting, and they must be revisited and viewed in a broader perspective of biological and sociocultural factors. The biological factors have to be stated in terms of the *primacy of the fight for survival and its universality, thus presenting sexuality as a second order issue greatly colored by the primary biological drive for survival.* The sociocultural factors must be related to the ever-changing morals and mores and to the inevitably limited empirical studies.

Power and Acceptance

The fight for survival is apparently the most general biological law. All living organisms fear death, but as far as we know, human beings are the only ones capable of imagining, thinking, and anticipating true or untrue dangers. With the exception of ants, squirrels, and a few other species, most animals seek food only when they are hungry and are unable to anticipate future shortages. No animals wage preventive wars or spend their entire life accumulating possessions they cannot possibly use, nor do any animals worry about remote or nonexisting threats to their lives.

Human beings seem to be the only species obsessed with the fear of not having enough to eat and/or the fear of falling prey to their true or imaginary enemies. Fear of dangers makes one wish to be able to overcome them. Thus, the strongest human motive is, necessarily, the desire to be strong. If survival is the arch-law of nature, the urge for power is its necessary corollary.

"Power" can be defined as the ability to defend oneself against enemies and to get food. Obviously the amount of power is measured by one's chances for survival. An organism is strong when it is endowed with tools and weapons which provide food and protect it against enemies; an organism is weak when it is poorly equipped. The peak of power is called omnipotence. An omnipotent being has all the power, that is, it can satisfy all its needs and ward off all threats. It cannot die of starvation nor be killed by its enemies. Obviously omnipotence necessarily includes immortality. Living organisms operate on a continuum of greater or lesser amounts of power; the less power the greater is the threat to their life; the more power they have, the better the chance for survival.

According to the interactional theory, the striving for power is a derivative from the general drive for survival, the *lust for life*. Most of human actions are a direct outcome or indirect derivative of the lust for life, and those actions which do not seem to be related to survival are colored by this fundamental drive. The care for one's health and physical strength and prowess is a clear expression of the lust for life. Accumulation of property, production of food, housing, and clothing, and the arms race reflect the drive for power.

Penis Envy

Medieval monks invented perverse, sadistic tortures for "women-witches" who were believed to have slept with the poor devils. The famous "Malleus" (witches' hammer) is a magnificant monument of projecting masculine perversions combined with saintly rationalizations. Freud's times were more tolerant of women, but Freud's contemporaries, such as A. Adler, believed that the lack of a penis was "organ inferiority."

It took Freud's genius to discover what others tried to hide. Many little girls wished that they were boys, for this was the only, though imaginary, way of escaping discrimination and subservience.

In the Victorian era, marriage was the only socially acceptable avenue for women. Unmarried women were ridiculed, ostracized, and blamed for remaining single. When a girl was unhappy being a girl and preferred an active and independent life pattern, she was called tomboy, amazon, masculine. To be feminine meant to oscillate between infantile dependence and motherly worries. When a woman refused to accept the three great feminine Ks—Küche, Kirche, Kinder (kitchen, church, and children)—she was treated as an outcast.

In Freud's time masculinity and femininity could have been described as follows:

> When you say 'masculine' you mean as a rule 'active,' and when you say 'feminine' you mean passive . . . The male sexual cell is active and mobile; it seeks out the female one, while the latter is stationary and waits passively. This behavior of the elementary organism of sex is more or less a model of the behavior of the individuals of each sex in sexual intercourse. The male pursues the female for the purpose of sexual unity, seizes her, and pushes his way into her. (Freud 1932:156)

Freud did not invent penis envy but discovered this widespread phenomenon in his times. The more restrictions were imposed on girls, the more frequent the wish to escape the yoke by a magic switching to the opposite sex.

Penis envy was never a general feeling common to all women at all times; certainly the Tschambuli or Arapesh women never had the reason for such an envy. In Arapesh, men and women shared household and child-rearing responsibilities, and the Tschambuli women were the dominating sex.

Freud's clinical observations of penis envy in women who were reared in an atmosphere of discrimination and subjugation must be interpreted in light of another hypothesis brought forward by Freud, namely the tendency of the child to identify with the "strong aggressor" (Fenichel 1945). In patriarchal families, the father was the absolute rule, and the male and female children were proud to identify with the father rather than the mother. All of Freud's writings stress the preference for a masculine father-based superego (Freud 1938).

One must, therefore, interpret penis envy in girls not as an envy of their little brothers or playmates but rather as a wish for the possession of the large fatherly penis and with it, the *fatherly power*. The penis, as the cherished symbol of power, was envied by women not because it was a sexual tool, for vaginas can undoubtedly procure as much and often more frequent sensual pleasure than penises; it was penis as the *power symbol* which elicited the well-understood envy feelings.

Sociopsychological Concepts

Sociopsychological studies (London 1977, Wolman 1958) led to a classification of interactional patterns according to the *aims* of their participants, depending on whether the main purpose of the interaction is the satisfaction of their own needs (instrumental), or their partners' needs (vectorial), or both (mutual or mutual acceptance).

When people enter a social relationship in order to have their needs satisfied and have in mind to take and not to give, it is an *instrumental* relationship. The individual regards others as tools or instruments. Human life starts as a parasitic process; the infant must receive all that is necessary for his survival. Thus the infant vs. mother relationship is the prototype of instrumentalism.

Mutual or mutual acceptance relationships develop gradually. Already in the nursery school or kindergarten the child is presented with opposition to his instrumentalism. The child learns to share, to take turns, to trade toy for toy, and to accept mutuality as a basis for interaction.

Mutuality achieves its peak in friendship and marriage. Each partner desires to make his partner happy and expects that the same is his partner's feeling. Successful marriage is based on mutuality; each marital partner is determined to do his best for the well-being of his or her partner and expects the same from the other party.

As long as one feels weak, he seeks help, but when one feels rich and strong, he is inclined to give away. Strong and secure men tend to be generous.

Parenthood is the prototype of *vectorialism*. Parents create life, protect it, and care for it, irrespective of their child's looks, health, I.Q., disposition, and success. The more the infant needs their help, the more sympathy he elicits.

Children could not make good parents; they demand love to be given to them; they are instrumental. So are childish adults—they cannot be adequate parents. To be an adequate parent one needs to feel strong and friendly and willing to give and to be ready to care without asking anything in return. To give without asking, or expecting anything in return is the highest degree of love and essence of vectorialism.

A normal and well-adjusted individual is balanced in social interaction. He is instrumental in the struggle for survival, mutual in relationships with friends and in marriage, and vectorial in regard to children and to those who need help. He is reasonably selfish (instrumental), reasonably mutual, and reasonably vectorial.

Mentally disturbed individuals cannot preserve this balance. They are either *hyperinstrumental,* displaying infantile selfishness and parasitism, or they neglect themselves and worry constantly about others in a morbid *hypervectorialism,* or they exaggerate in giving and in taking, in shifting moods of *dysmutualism.*

This division into three types of interindividual relations can be made, isomorphic to the psychoanalytic theory, provided Freud's concepts of cathexis be applied to interindividual processes. A new concept must be introduced: the "interindividual cathexis." There are no neurological or physiological counterparts to this concept. It may be analogous to Pavlov's explanation of reflex. An external stimulus, wrote Pavlov, is "transformed into a nervous process and transmitted along a circuitous route (from the peripheral endings of the centripetal nerve, along its fibers to the apparatus of the central nervous system, and out along the centrifugal path until, reaching one or another organ, it excites its activity)" (1928: 121). Pavlov's description can be explained in terms of cathexis of physical energy; the external stimulus transmits a part of its energetic load into the peripheral endings of the centripetal nerve, and thus cathects, charges this nerve ending, and through the circuitous route it cathects the nerve center.

There is no evidence that mental energy is cathected and that cathexis follows the analogy with the physical energy. The term *interindividual cathexis* is introduced as a logical construct because of its methodological flexibility and usefulness. The concept of interindividual cathexis permits the building of a bridge between psychoanalytic studies of personality and experimental research in social psychology. When A loves B, A's libido is object-cathected in B (Wolman 1966b).

In instrumental relationships the individual's own libido is self-cathected and the individual expects the libido of others to be object-cathected into him. The infant attitude is instrumental, i.e., he wishes to receive cathexes, to become a libido-cathected object. The same applies to any other instrumental relationship.

In mutual relationships the individual aims at receiving libido cathexes as well as giving them to those from whom he aims to receive. It is a give-and-take relationship. Satisfactory sexual intercourse is usually such a relationship. So is true friendship; each partner desires to object-cathect his libido (aim-inhibited, if it is a nonsexual relationship) in his partner and expects his partner to have the same aim.

In vectorial relationships the individual aims to object-cathect, to give libido to others. Parenthood is a prototype of such a giving without receiving, a one-way object-libido cathexis. It is, as a rule, an aim-inhibited libido cathexis.

Mature and well-adjusted people are capable of functioning in all three types of relationships. They are adequately instrumental to protect themselves and to satisfy their own needs. They are adequately mutual in sexual and nonsexual friendship relations. They are adequately vectorial in parenthood and in their charitable and idealistic deeds. In psychoanalytic terms, the libido is reasonably balanced in interindividual relationships and properly divided between inter- and intracathexes; their destrudo is partially sublimated and partially kept under rational control.

A dysbalance in interindividual cathexes, caused by interaction between the individual and his environment, inevitably leads to a dysbalance in intraindividual cathexes. In other words, *improper social relations in early childhood must cause personality disorders*. Thus all sociogenic disorders can be divided into three types: the hypervectorial object-hypercathected, schizotype disorders; the hyperinstrumental, self-hypercathected, psychopathic type; and the dysmutual, going from one extreme to the other, cyclic type.

Interindividual Cathexis

The development of adjustive and maladjustive personality traits is brought about by a variety of factors, among them heredity, maturation, learning, and sociocultural influences. The interactional theory stresses the latter, and introduces the concept of interindividual cathexis (Wolman 1966a: 24 ff.).

When Freud wrote about object relations, it was about the individual who cathects and not about the cathected object. When A loves B, said Freud, A's libido is cathected. But what happens to B when A loves him? Or when love is mutual? Freud dealt only with the distribution of the libido *within* the individual. Narcissism is the cathexis of the libido in oneself, object-relation is the cathexis of libido in others. But what happens to those toward whom the cathected libido turns is a matter that requires further study.

The sociologically oriented neoanalytic schools of Horney and Sullivan took up this issue. Horney devoted a great deal of attention to the *need to be loved.* When a child feels loved, accepted, and cared for, he develops a feeling of *safety.* When the child feels rejected, a *basic anxiety* develops. Human activities are guided by both pleasure and safety; people would renounce pleasure rather than safety, said Horney (1939: 73). Instead of Freudian's sexual or aim-inhibited love, she introduced an environmentalistic concept of protection and safety. Also, instead of Freud's active cathexis of libido or need to love, she emphasized the need to be loved.

Horney and Sullivan added important aspects to the study of social relations. Horney emphasized the "receiving object," Sullivan the interpersonal involvement. Yet there is no need to reject Freud's model of personality and substitute it by the models introduced by Horney or Sullivan. The comparison between these three theories of personality (Wolman 1981a, ch. 6 and 9) would show the superiority of the Freudian model of personality. Freud's model of personality can be adapted to interindividual relations by a broadening of Freud's term cathexis.

The concept of interindividual cathexis is a *theoretical construct.* This construct is introduced because of its methodological flexibility and usefulness. The term interindividual cathexis allows one to make full use of Freud's model of personality that offers so many methodological advantages and that has been tested in the vast experience of psychoanalytic therapy. While being faithful to Freud's principles, we are able to understand the interpersonal relations better, for the cathected object is included in our studies. Furthermore, the concept of interindividual cathexis permits us to build a bridge between the body of psychoanalytic studies and experimental research in social psychology in terms of the theory of power and acceptance.

Consider the instrumental type of relationship. It is a "getting" type of relationship. The individual's aim is to receive libido cathexes from others. His libido is self-cathected and he expects the libido of others to be object-cathected into him. An infant wants to be loved, or accepted (Horney), or approved (Sullivan). In accordance with Freud's personality model, the infant wishes to receive cathexes and to become a libido-cathected object. The same applies to any other instrumental relationship, such as the desire to be supported, to be popular, and to be appreciated and admired.

In a mutual type of relationship, the individual wishes to receive libido cathexes as well as to give them to those from whom he desires to receive. It is a give-and-take relationship. So is true friendship; each partner desires to object-cathect his libido (aim-inhibited, if it is nonsexual relationship) in his partner and expects his partner to have the same aim.

In a vectorial type of relationship, the individual aims to invest, to object-cathect, to give to others. Parenthood is the prototype of such a giving without receiving, one-way object-libido cathexis. It is an aim-inhibited cathexis, of course. So is psychotherapy. But whatever exists does so in a certain quantity. Bettelheim noticed that when the emotional resources of workers who worked with psychotic children were drained, these workers needed to "receive replenishment of their libido through contacts with others and through narcissistic satisfactions" (Bettelheim 1955: 170).

Freud's developmental stages describe object relations viewed from the vantage point of the developing child. The interactional theory adds to the picture the interaction between the child and the parents.

The child's object relations include cathexis of the libido directed toward others (giving) and their cathexes directed toward the child. Thus *parental attitudes* become a crucial factor in shaping the child's wholesome or pathological personality development.

Developmental Stages

From the beginning of the intra- and later extrauterine life, the new organism is a taker. One may hypothesize that all the available libido of the neonate is self-cathected, and all his destrudo is directed against the outer world for self-defense.

The higher the species, the longer and more complex is the process of growth and maturation. In humans, birth is the beginning of a prolonged maturational process that leads from a wholly parasitic beginning toward a nonparasitic maturity. Maturity includes the ability to support oneself economically, to perform sexually, to assume parental responsibility, and to cooperate with one's social environment.

Kardiner (1945) related personality differences to child-rearing practices. Whiting and Child (1953) and others have shed additional light on the impact of sociocultural factors upon the child's growth and development. The tendency to pass through certain developmental stages is universal as demonstrated by the work of many researchers, including Piaget and Gesell. This research basically affirms Freud's concept of psychosexual developmental phases. However, it is apparent that the way in which a child passes through the developmental stages and the resulting degree of fixation and regression depends largely on what the individual child inherited and the child's *interaction* with his particular environment. When a child is exposed to a friendly, wholesome, and affectionate environment, his libido, initially self-directed, is lured into object cathexes. In an overdemanding environment, there is a danger of object hypercathexis and self-hypocathexis (Wolman 1957; 1981). In a rejecting and hostile environment, object cathexis of destrudo is encouraged.

Threat produces *fear;* fear is the alarm bell of the organism. Fear produces either fight or flight or both. Those who have no fear cannot be hostile. If absolute security were possible, fights and wars would disappear. People fight either as greedy aggressors who fear they may die unless they devour someone, or as defenders who fear death inflicted by aggressors. Those who fear, hate and those who hate, fear.

In neonates, neither reaction is tempered by any significant control apparatus. There is no need to reiterate here the Hartmann, Kris, and Loewenstein (1946) theory of ego; if there is any rudimentary ego, it is too weak to inhibit wild outbursts of destrudo. Fear in infants means panic; fight means furious rage. In autistic children and in catatonic adult schizophrenics (Wolman 1966a, 1968, 1970) there is no calculated evasion of risk or planned escape, nor is there a goal-directed attack on enemies; instead, there is blind panic or wild rage, often senseless and indiscriminate.

Under normal circumstances, libido becomes gradually and to a great extent object cathected. In the normal adult libido is distributed between self and object cathexes which enables him both to take care of himself and to be friendly to others.

There are, however, no developmental stages of destrudo. Destrudo is archaic, primordial, primary. There can be no developmental stages in destrudo analogous to Freud's developmental stages of libido, because destrudo cannot progress. Destrudo serves self-defense and survival; when object love fails, part of frustrated libido turns into secondary narcissistic love, and part into destrudo. Probably destrudo and libido are, as it were, two levels of the same mental energy, two kinds of fuel that are transformable into each other. In states of emergency, severe deprivation, lack of sleep and exhaustion, the destrudo takes over. Destrudo, as the more primitive energy of the organism, is always present and leads to outbursts of rage in animals and to temper tantrums in infants.

In terms of the power and acceptance theory the oral stage is an early form of *instrumentalism*. It is an interindividual relationship in which the desire to be helped (to get) is combined with considerable fear and hostility. The "weak" infant needs food and must receive it at any cost. When refused, all four forms of hostility (i.e, aggression, defense, panic, and horror) are promptly elicited.

The anal stage is characterized not only by toilet training but also by walking and talking behavior. The erect position and walking around permit the child to use hands and manipulate objects to a much greater extent than he could have done in the oral stage. The child can explore distant places, disappear from mother's sight, and reach into new areas hitherto forbidden.

The beginning of speech is another source of ever-increasing power. To be able to comprehend and be comprehended, to call and be called to, to exchange wishes and demands—all this gives the infant a new feeling of power. Perhaps infants hallucinate power, but the toddlers who babble nonstop obviously enjoy their torrent of talk.

Another source of power lies in toilet training. At the anal stage, unless the mother is sadistic and resorts to enemas, she is often at the mercy of the child. The mother can seat the child on the toilet, but thereafter she relinquishes her control. Mother may refuse food to her child, but feces are the child's indisputable possession, the prototype of money and of any future property.

At the phallic stage, object love is channeled into genital desires directed toward the parent of the opposite sex, and becomes, for the first time, the dominating force, stronger than destrudo. At the oral stage, the cannibalistic object relationship (unfortunately called cannibalistic "love") is dominated by destrudo. Infants eat, and eating is destruction. When one says, "I love chicken," he really does not love it; he merely "loves" to eat it, and by eating it he performs an act of destruction. There is no object love at the oral stage; the only love is self-love, that is, self-directed libido.

Cathexis and conditioning are intertwined in this developmental process. In daily interaction with other children, the child clings to his own possessions while trying to take hold of the attractive toys his playmates display. The playground, the sandbox, the backyard, and the nursery school all serve to teach the child mutuality, for children are instrumental-minded and selfish. They will typically engage in numerous frustrating confrontations until, with the guidance given by adults, they learn to give in order to receive, and to share in order to enjoy, thus taking the first steps in the direction of *mutualism*.

Hostility

The Oedipal conflict signals future cooperation and competition among adult members of the same sex. Under normal circumstances, the Oedipal conflict leads to inhibition of love and hate. Hate can be inhibited either by love or by fear, or by a combination of both. The strong and friendly father evokes fear by his strength, and love by his kindness. A fusion of love and fear gives rise to the feeling of *awe* and thereby leads to development of the superego.

Freud believed that Thanatos would eventuate in one's own destruction unless it were diverted into overt hostility against the outer world. Part of this hostility, directed against the restricting and prohibiting par ents, becomes incorporated in the superego and channeled against one's own ego, that is, against oneself. It is my contention that the destrudo serves survival and that it is initially directed against threats from without.

The fight for survival is better served by a rational, controlled and goal-directed hostility. The ego, operating on the "reality principle," delays and controls destrudo reactions. Instead of reacting with an outburst of the panic or terror variety, the ego exercises judgment. The reality principle is not an antipleasure rule; rather it is a rule given to pursuit of pleasure at minimum risk. Whereas the id is prone to fight at the slightest provocation, the ego fights to win. Thus, the ego-seated discharges of destrudo are aggressive or defensive or both, always aiming at the well-being of the organism.

With the formation of the superego, a part of the destrudo becomes invested therein. The superego is formed through the introjection of both parental prohibitions and the idealized parental image. The former carries some elements of destrudo; the latter includes elements of libido. When the superego is first set up, it is invested with some of the child's aggressiveness against his parents. This hate can hardly find satisfactory outlets; normal children do not kill nor do they attack the parent of the same sex. Thus, the thwarted destrudo turns inwardly, against the ego, creating feelings of guilt.

During the time that Freud lived, the latency period was the period of life when inhibitory forces acted upon the child's libido and destrudo. The child was turning away from the parent of the opposite sex and becoming antagonistic to the opposite sex in general. He still loved the parent of the opposite sex, but his love became desexualized and aim-inhibited.

Freud's formulations require a cultural-historical adjustment. In a society based upon rather stable interindividual relationships, such as was true of Vienna at the beginning of this century, boys renounced mothers as love objects and girls renounced fathers. Boys played with boys, girls with girls; boys played cops and robbers; girls played house. Boys avoided girls and girls avoided boys. Eros became aim-inhibited, love de-emotionalized, and groups were formed based on hierarchic social structure, tight and exclusive of the opposite sex.

There is, today, much less inhibition and practically no delay in sexual information compared to that existing in Freud's times. Thus, it seems that the time span of the latency period has shrunk, and not all children go through this phase in the way Freud described it. There is no doubt that the differences in child rearing strongly affect developmental problems and the ultimate personality structure (Wolman 1981).

Thus, the developmental theories of Freud, Piaget, Gesell, Erikson, and others require some sociocultural correction. These theories certainly hold true for those cultural groups where parental authority and especially father's authority survived two world wars, social upheavals, mass communication, and rapid social mobility. In these more conservative and traditional social groups, latency is still the period of aim-inhibited love for the parent of the opposite sex and aim-inhibited hate for the parent of the same sex. The future will tell to what extent development stages will be affected by sociocultural changes (cf. Kardiner 1945; Erikson 1963; Whiting and Child 1953). The fact is that in some primitive tribes, such as the Marquesans and Trobrianders, where there is little if any cultural restraint, sex is practiced in preadolescent years and the concept of latency is inapplicable to their developmental years.

Children, at the middle childhood or latency stages, are far from having achieved self-sacrificing vectorialism. They are hardly capable of a genuine give-and-take mutualism. Wolman's (1951) study of spontaneous groups and Piaget's (1948) classic study of moral judgment in children point to the lack of genuine compassion, sympathy, and consideration at this age. Children seek friends not in order to help but to be helped. Whenever they reciprocate, it is not because of their heart's innermost desire, but because otherwise they would lose their friends. Be that as it may, this unwilling reciprocation is a step forward toward mutuality.

The biological changes in adolescence seriously affect the balance of the intra- and interindividual cathexes. Puberty is, undoubtedly, a period of substantial changes, for it is at this point that the adolescent becomes capable of reproduction and ultimately attains biological adulthood. In several primitive societies puberty rites lead to the onset of the three most significant functions of adulthood, namely breadwinning, marriage, and participation in community life. In civilized societies physical maturity antedates psychological and cultural maturity, creating a situation that breeds conflict.

Moral Development

My theory proposes that moral development goes through five phases, namely, anomy, phobonomy, heteronomy, socionomy, and autonomy, (Wolman 1976). According to this theory, people are born *anomous*. They know of no restraint, no concern, and no consideration for anyone. The intrauterine life is clearly parasitic: the zygote-embryo-fetus is all out for himself and grabs whatever he can. Mother's body is self-sacrificing: it is ready to give whatever it possesses to feed and to protect the not-yet-born child.

Birth changes little in the parasitic attitude. Although the newborn must use his or her own respiratory system, the parasitic-dependent attitude continues even after cutting the umbilical cord. The infant wants what he wants and follows his instant stimuli and impulses. He operates on the principle of immediate gratification of needs called by Freud *Lustprinzip* (erroneously translated as pleasure principle). Infants are anomous, that is amoral and limitlessly selfish.

Morality means concern for fellow men, but it starts with restraints imposed from without. Fear of retaliation and punishment is the first, though preciously small, step toward behavior. The earliest restraints and rules are based on fear, thus the earliest phase of primitive morality should be called *phobonomous*. With the development of rudimentary awareness of potential consequences, called by Freud "archaic ego," the child's selfishness faces restraints. He learns to obey because he fears. Fear is not morality, but primitive people and toddlers must be restrained from without. This restraint breaks the ground for more advanced phases of morality.

As the child becomes aware of parental love and care, he appreciates what he gets and begins to reciprocate. His "love" is quite selfish, for he loves only those who love him. He wishes to protect them, for he needs their protection. The child needs parental love and fears he may lose it, thus his attitude toward the parents is a combination of love and fear. The child willingly and fearfully accepts parental rules and prohibitions, and gradually absorbs these rules and perceives them as if they were his own. He may identify with the parents or parental substitutes and incorporate their prohibitions and norms. He may blame himself for occasional disobedience and devolop *guilt feeling* whenever he violates parental rules. His behavior is *heteronomous*, for he willingly obeys norms instituted by others whom he perceives as loving and powerful authority figures. In Freud's personality model this self-regulatory agency was called *superego*.

Moral development does not reach its full development with a childhood acceptance of parental rules and the formation of superego. Preadolescents and adolescents develop close interpersonal relations with their peers and form groups, cliques, and gangs. Quite often these new social relations displace the child-parent attachments, and the rules of peer society become the ultimate source of moral or antimoral behavior. The individual's willful identification with the peer group to which he chose to belong becomes the guiding principle in his life for years to come, and often his loyalty goes on until the end of his life. This willing acceptance of social norms, be it of certain religious denomination, racial or ethnic group, or a political party leads to the formation of a new part of one's personality, which I would call the *we-ego,* and this group-identification period I shall call *socionomous*.

Most people do not go beyond the socionomy and acceptance of the norms prevailing in their particular group, clan, subculture, or of some larger social segments. Most people follow certain rather limited moral rules binding in their particular group. They are "brothers" and "sisters" who abide by moral principles within their brotherhood and sisterhood.

The ultimate moral development goes beyond socionomy. It is related to the idea of power. Imagine an Omnipotent Being. An Omnipotent Being does not fear and does not need anything, thus He cannot hate nor hurt anyone. The only thing He can do is to give what He has. He *must* build and create, love and protect. Those who reached this divine point, as Christ, Prometheus, Antigone, and some genuinely moral individuals developed a vector-ego, an ego that goes out of themselves and reaches toward others. They have the *courage to give*. They do not need any restraint, for their life is devoted to helping others. *Summa ethica* is nothing else but *summus amor*. They are autonomous.

BIBLIOGRAPHY

Bettelheim, B. 1955. *Love Is Not Enough*. Glencoe, Ill: Free Press.

Dollard, J., N. E. Miller, L. W. Doob, O. H. Mowrer, and R. R. Sears. 1939. *Frustration and Aggression*. New Haven: Yale University Press.

Erikson, E. 1963. *Childhood and Society*. Rev. ed. New York: Norton.

Fenichel, O. 1945. *The Psychoanalytic Theory of Neurosis*. New York: Norton.

Freud, S. Standard Edition. James Strachey, ed. 24 vols. London: Hogarth, 1962; New York: Macmillan, 1962.

——1898. Sexuality in the etiology of neurosis. Standard Edition, 1: 263–286.

——1905. Three essays on the theory of sexuality. Standard Edition, 7: 130–245.

——1931. Female sexuality. Standard Edition, 21: 225–243.

——1933. *New Introductory Lectures on Psychoanalysis*. New York: Norton.

——1938. *An Outline of Psychoanalysis*. New York: Norton, 1949.

Hartmann, H., E. Kris, and R. M. Loewenstein. 1946. Comments on the formation of psychic structure. *The Psychoanalytic Study of the Child*, 2: 11–38. New York: International Universities Press.

Horney, K. 1939. *New Ways in Psychoanalysis*, New York: Norton.

Kardiner, A. 1945. *The Psychological Frontiers of Society*. New York: Columbia University Press.

Klein, G. S. 1965. *Essays on Ego Psychology*. New York: International Universities Press.

Lawson, R. 1965. *Frustration*. New York: Macmillan.

London, H. 1977. Power and acceptance theory. In B. B. Wolman, ed., *International Encyclopedia of Psychiatry, Psychology, Psychoanalysis, and Neurology*, 9:11—15. New York: Aesculapius.

Pavlov, I. P. 1928. *Lectures on Conditioned Reflexes*. New York: Liveright.

Piaget, J. 1948. *The Moral Judgment of the Child*. New York: Basic Books.

Whiting, J. W. and I. L. Child. 1953. *Child Training and Personality*. New Haven: Yale University Press.

Wolman, B. B. 1951. Spontaneous groups in childhood and adolescence. *Journal of Social Psychology*, 34: 171–182.
——1957. Explorations in latent schizophrenia. *American Journal of Psychotherapy*, 11: 560–588.
——1958. Instrumental, mutual acceptance, and vectorial groups. *Acta Sociologica*, 3: 19–28.
——1966a. *Vectoriasis Praecox or the Group of Schizophrenias*. Springfield, Ill.: Thomas.
——1966b. Transference and countertransference as interindividual cathexis. *Psychoanalytic Review*, 53: 91–101.
——1970. *Children Without Childhood*. New York: Grune and Stratton.
——1976. Moral principles. *International Journal of Group Tensions*, 6: 1–4.
——1981. Interactional theory. In B. B. Wolman, ed., *Handbook of Developmental Psychology*. Englewood Cliffs, N.J.: Prentice Hall.
—— 1981a. *Contemporary Theories and Systems in Psychology*. New York: Plenum.
——*Being—Lasting—Becoming*. (In press).
Woman, B. B. and J. Money 1980. *Handbook of Human Sexuality*. Englewood Cliffs N.J.: Prentice Hall.

[PART IV]
NON-FREUDIAN THEORIES

[FOURTEEN]
Adler's Individual Psychology

Adler's theory was built on the foundations of Freud's theory of unconscious motivation, Dilthey's psychology of understanding, Smut's holism, and Vaihinger's fictionalism. His theory is a combination of organismic and sociological interpretation of human behavior.

In the years 1911–1912 Freud's theory was formulated in terms of libido–sex and ego–self-preservation instincts. The theory of the death instinct, of the superego, and the further development of the theory of ego, anxiety, and defense mechanisms came later. Adler's rebellion started with the problem: What is the driving power in man? Is it the hedonistic libido? Or the self-preservation instinct? And in all these questions Adler's answer was quite different from Freud's.

According to Freud, social development is a function of libido development. Adler was opposed to this idea. He wrote that it would not be possible to maintain that every drive has a sexual component. Adler separated sociability and sexuality. The social interest or sociability is "an innate potentiality which has to be consciously developed."

Adler rejected Freud's ideas of socialization of the child through repression, reaction formation, aim inhibition, envy, and superego formation. Humans are born with the "innate potentiality." The Oedipus complex, is, according to Adler, not a universal pattern of development but a product of erroneous upbringing of children.

Adler believed that the child's initial and innate impulses of affection were directed toward others and not toward himself. Freud believed that the newborn child was a narcissist; Adler believed that self-boundedness is an artifact imposed upon the child by his education and by the present state of our social structure. The creative power of the child is misled toward self-boundedness instead of being guided into the normal channels of cooperation with other individuals.

While Freud's impact on Adler's clinical studies cannot be doubted, there is the undeniable fact that Freud's and Adler's philosophical concepts are almost diametrically opposed. Thus, the aim of these lines is to highlight some of the differences in the *Weltanschauung* of these two thinkers.

Freud was all his life a biologically oriented scholar who had never abandoned the framework of natural sciences. Freud stuck to certain basic concepts of the natural sciences even when his writings reflected the impact of Brentano, Schopenhauer, or Herbart.

While Freud was never orthodox in regard to his own concepts and modified them whenever it was deemed necessary as his empirical findings and theoretical concepts progressed, he never gave up the following three principles, generally accepted by natural scientists at the crossroads of the nineteenth and twentieth centuries: namely causality, reductionism, and homeostasis.

Charles Darwin's theory of evolution and adjustment, Ernst Haeckel's principle of biogenetics, the theories of preservation of energy and homeostasis, and a strict determinism form the bases of psychoanalysis. Although Freud formulated a psychogenic theory of mental disorder in contradistinction to neurological theories, he postulated that mental energy was a sort of derivative of the energy that activates all living substance.

The Creative Drive: Some Theories

Adler leaned heavily on the *Kulturwissenschaft* and believed men were driven by "creative power." It was a purely human force, not related to biological factors. Adler, like all "understanding" psychologists, believed in the uniqueness of human nature and in the uniqueness of each individual. In the division of the sciences into natural and cultural sciences, as proposed by Windelband, Freud belongs to the first category, Adler to the second.

Adler wrote:

> Types, similarities, and approximate likenesses are often either entities that owe their existence merely to the poverty of our language, which is incapable of giving simple expression to the nuances that are always present, or they are results of a statistical probability. The evidence of their existence should never be allowed to degenerate into the setting up of a fixed rule. Such evidence cannot bring us nearer to the understanding of the individual case. It can only by used to throw light on a field of vision in which the individual case in its uniqueness must be found. (Ansbacher and Ansbacher 1956: p. 193)

Kulturwissenschaft

The nineteenth-century neo-Kantians, and especially Wilhelm Windelband and Heinrich Rickert, have exercised considerable influence in several circles of the academic world. One must remember that in those times even medical students were supposed to study philosophy, and both Freud and Adler have been influenced by contemporary philosophical treatises. While Freud was more influenced by the materialistic-mechanistic way of thinking, Adler came closer to the neo-Kantians.

Windelband and Rickert counterposed the cultural or historical sciences *(Kultur* or *Geschichtswissenschaften)* to the natural sciences. They preferred the name "cultural sciences" to the old usage of "humanities" *(Geisteswissenschaft)*, for "humanities" included the eternal values of logic, ethics, and aesthetics, while the *Kulturwissenschaft,* so they believed, dealt with empirical and passing events and values.

Both *Naturwissenschaft* and *Kulturwissenschaft* are empirical sciences. They differ in methods of research. The natural sciences seek the principles and laws that govern the given subject matter. Natural sciences are *nomothetic,* i.e., they seek general laws. The natural sciences are not interested in the single case; they are concerned with the necessary occurrences, with the laws that govern the universe, while the *Geisteswissenschaften* deal with the individual, the unrepeatable and unique case; e.g., it happened just once that Napoleon retreated from Moscow in 1812. This event was unique and unrepeatable; it was an *idiophenomenon.* Sciences dealing with idiophenomena do not seek general laws; the idiographic sciences are concerned with values and not with natural causes.

A similar cleavage between natural sciences and humanities was introduced by W. Dilthey (1833–1911). Dilthey was strongly opposed to the mechanistic-deterministic philosophy and espoused vitalism, later on developed by Henri Bergson. Following Kant's distinction between the ultimately unknowable "natural objects" and the "immediately given" experience, Dilthey wrote in 1884: We know natural objects from without through our senses. However we may break them up or divide them, we never reach their ultimate elements in this way. . . . How different is the way in which mental life is given to us! In contrast to external perception, inner perception rests upon awareness *(Innewerden)*, a lived experience *(Erleben)*, it is immediately given" (Hodges 1949: 133).

In his opus magnum, called *Ideen über eine beschreibende und zergliedernde Psychologie* (Ideas Concerning a Descriptive and Analytical Psychology), Dilthey stressed the deep difference between psychology and natural sciences. The task of the natural sciences is to *explain (erklären)*, while the task of psychology is to *understand (verstehen)*. In understanding we start from the system as a whole, which is given to us as a living reality *(der uns lebending gegeben ist)*, to make the particular intelligible to ourselves in terms of it (Hodges 1949: 136).

Holism and Purposivism

The holistic approach is the underlying factor in E. Spranger's and W. Stern's personalism, Wertheimer's Gestalt, K. Lewin's field theory, and A. Adler's individual psychology. "The same tones tell a different tale in Richard Wagner and in Liszt," wrote Adler (1931: 403).

Following in the footsteps of Windelband, Adler assumed the uniqueness and unrepeatability of mental phenomena. This approach brought Adler's theory close to Gestalt and personalistic psychologies.

In accordance with Windelband's philosophy, Adler assumed that events in nature "are causally determined; but in psychology we cannot speak of causality or determinism." In psychology, Adler wrote, "We regard man as if nothing in his life were causally determined and as if every phenomenon could have been different." And further on: "Every semblance of causality in the psychological life is due to the tendency of many psychologists to present their dogmas disguised in mechanical or physical similes" (Ansbacher and Ansbacher 1956: 91–92).

According to Adler, "Alles kann auch anders sein" (Everything may turn out differently; 1933b: 7). Events are not determined by their causes, but directed by their goals. The history of mankind was not determined by a cause-effect relationship but by human strivings toward goals.

Adler suggested substitute causation by purposivism. "All psychic activies are given a direction, a previously determined goal. . . . Every psychic phenomenon, if it is to give us any understanding of a person, can only be grasped and understood if regarded as a preparation for some goal" (1928: 4). The main goals, as it will be later explained, are the strivings toward perfection, toward a feeling of superiority.

"As If" Philosophy

In 1911 Hans Vaihinger published a major philosophical work called *Die Philosophie des Als-Ob* (The Philosophy of As If). In opposition to the prevailing positivistic-materialistic climate, Vaihinger defined his theory as "idealistic positivism" or "positivistic idealism."

Vaihinger's position is best defined as "fictionalism." Vaihinger distinguished between hypotheses and fictions. A hypothesis is something that may be eventually proven and established as truth; a fiction is a "mere auxiliary construct," "a scaffolding to be demolished" (Vaihinger 1924: 163).

Adler applied Vaihinger's principles toward the interpretation of human behavior. Adler did not maintain that the neonates are endowed with a clearly established drive toward superiority. They act, however, *as if* they were led by this goal. Adler wrote:

> The fictitious goal, vague and labile, not capable of being measured, constructed with very inadequate and ungifted powers (of the infant) has *no* real existence, and can therefore not fully be comprehended causally. It may, however, be understood as a teleological, artifical device *(Kunstgriff)* of the mind, which seeks orientation and which, in emergencies, is always constructed concretely. (1930: 4)

Every child is born with a free creative power. This innate power is exposed to influences stemming from the child's organism and his environment. As a result of the interaction between his free creative power and these factors, the child chooses a style of life. The free creative power is no longer free; it becomes involved in the total structure of the individual personality. It is the main determinant of personality because it utilizes the individual's heredity and environment.

Thus the decisive factor in life is the *law of movement*. Adler believed that the establishment of the law of movement was perhaps the greatest step in the development of his theory. This law indicates the way the individual arrives at the solution of his problems and the manner in which he overcomes the obstacles in his life. The law of movement is individual and unique for each person; each individual can be characterized and recognized by his law of movement.

Inferiority Complex

In a paper published in 1910, Adler wrote:

> For children with inferior organs and glandular systems . . . their growth and functioning shows deficiencies and that sickness and weakness are prominent especially in the beginning. . . . These objective phenomena frequently give rise to a subjective feeling of inferiority. . . . Such children are thus often placed in a role which appears to them unmanly. . . . The renunciation of masculinity, however, appears to the child as synonymous with femininity. . . . Any form of uninhibited aggression, activity, potency, power, and the traits of being brave, free, rich, ag-

> gressive, or sadistic can be considered as masculine. All inhibitions and deficiencies, as well as cowardliness, obedience, poverty and similar traits, can be considered feminine.

The more intense the deprivation, the stronger the urge for compensation. The entire life of the child, his dreams, wishes, and play are filled with his goal to be a man, to be big and strong. Hence the "masculine protest."

Both concepts, the inferiority and the protest, underwent changes in Adler's writings. In 1933 Adler emphatically stated that "to be human means to feel inferior," for every child is inferior in the face of life and of adults. The weakness of the child, his ever-present inferiority feeling now became the main problem in Adler's theory.

Adler gradually transformed the masculine protest into a general striving for adequacy *(Vollwertigkeit)*, for meaning something in life *(Geltungstrieb)*, for security. The child is born weak and within his first four or five years a goal is set. This goal is compensatory, and the striving toward it is a striving to "security, power, and perfection." It is no longer striving for sheer power but for *overcoming* the difficulties of life and of one's own inferiority feelings.

Inferiority-Superiority

Life is movement directed toward a better adaptation to life. According to Adler, "the compulsion to accomplish a better adaptation can never end; in this lies the basis for our concept of '*striving to superiority*' "(1933: 259).

Adler's theory of inferiority started with his observation of "organ inferiority," related to women's lack of a penis. Later on Adler generalized his findings and noticed that any bodily inadequacy or handicap, and any unfortunate social status or any failure in one's endeavors may give rise to a profound feeling of inferiority. In most general terms Adler's theory implies that whenever an individual fails in his striving for superiority, he experiences the *feeling of inferiority*. This feeling activates him to overcome this unbearable feeling, and develop compensatory ideals which will eventually restore the feeling of superiority. As Kurt Adler put it:

> In the unerring and single-minded pursuit of his goal, the individual will put all body functions, emotions, thoughts, perceptions, and the total pattern of his behavior into its service. The result is a unified striving, and as a consequence, the personality will itself be a self-consistent,

> unified entity. Apparent ambivalences or contradictions found in a personality or in the individual's behavior are simply different methods used in the pursuit of a single goal.
>
> For the understanding of a personality, we can therefore say: If we recognize the goal of a psychic expression or its life plan, we must expect that all partial expressions will be in harmony with it, and vice versa: correctly understood partial expressions must in their connections and totality add up to the picture of a unified life plan and its final goal.
>
> Insofar as they all strive for perfection, for overcoming, for superiority, the final goal of all people is the same. However, each individual's conception in regard to the *meaning* of perfection, overcoming, superiority, is *unique*. Each individual has, after all, constructed his final goal as a compensation for *his* sense of inferiority regarding *his* body, *his* advantages and disadvantages in his social relations. The personality is therefore a self-consistent, unified, and *unique* unit. An individual's mode of movement throughout life toward his goal is called his pattern or *style of life*. (1967: 300)

Adler noticed that the great orators Demosthenes and Demoulins stuttered in their early years, and the great composer Beethoven suffered ear disease, and a great many painters had some sort of defects of vision. Minority groups try to overcompensate for their inferior social status, and homely children often excel in their studies.

> Whether a person desires to be an artist, the first in his profession, or a tyrant in his home, to hold converse with God or humiliate other people; whether he regards his suffering as the most important thing to which everyone must show obeisance, whether he is chasing after unattainable ideals or old deities, over-stepping all limits and norms, at every part of his life he is guided and spurred on by his longing for superiority, the thought of his godlikeness, the belief in his special magic power. In his love he desires to experience his power over his partner. In his purely optional choice of profession the goal . . . manifests itself in all sorts of exaggerated anticipations and fears. (Adler 1930: 390)

Aggression and Sociability

Adler postulated two basic drives, aggression and sociability. Initially he believed that aggressive striving for self-assertion was the main driving force. He wrote in 1908:

> From early childhood, we can say from the first day (first cry), we find a stand of the child toward the environment which cannot be called

> anything but hostile. If one looks for the cause of this position, one finds it determined by the difficulty of affording satisfaction for the organ. This circumstance as well as the further relationships of the hostile, belligerent position of the individual toward the environment indicate a drive toward fighting for satisfaction which I shall call "aggressive drive." . . . Fighting, wrestling, beating, biting, and cruelties show the aggression drive in its pure form. (Adler 1956: 34–35)

But man was never alone, and group living is just as fundamental a fact of life as aggressiveness. Freud assumed that the human neonate is a narcissistic creature and his entire libido is invested in himself. Melanie Klein went even farther in assuming a primary hostile attitude in infants.

Adler rejected both the ontogenetic theory of primary narcissism and the phylogenetic theory of socialization by restraint from without. He wrote:

> . . . when other schools of psychology maintain that the child comes into the world a complete egoist with a "drive for destruction" and no other intention than to foster himself cannibalistically on his mother, this is an erroneous inference based on incomplete observation. . . . These schools overlook in the relationship the role of the mother who requires the cooperation of the child. The mother with her milk-filled breasts and all the other altered functions of her body . . . needs the child as the child needs her. (1956: 129)

Adler quoted Darwin, who

> already pointed out that one never finds weak animals living alone. Man must be included among these, particularly because he is not strong enough to live alone. He has only little resistance against nature, he needs a larger amount of aids to live and preserve himself. . . . Man could maintain himself only when he placed himself under particularly favorable conditions. These, however, were afforded to him only by group life. *(Ibid.)*

Sociability or social feeling *(Gemeinschaftsgefühl)* is one of the key concepts in Adler's system. One wonders why Adlerian psychology carries the name "individual psychology" when it is a *par excellence* a social psychology. Psychologists are unable to comprehend any human experience outside the scope of social relations, Adler wrote. The conscious, or consciousness itself, is a product of social interaction. The universal striving for self-assertion and superiority is tempered by the social feeling, and the inner harmony between the individual's drive for mastery and his consideration for fellow men makes for a wholesome and mentally healthy personality.

Freud, too, was aware of the impact of environment upon personality structure. The instinctual innate forces and the developmental stages are significant factors, yet the final outcome depends a great deal upon the individual's interaction with his environment. Fixations and regressions, formations and malformations of the ego and the superego depend upon our environment.

Adler presented the issue in a different light. Freud perceived the individual as a separate entity exposed to the restraining social forces. Adler emphasized that community precedes the individual life. Man has always been in a society, never outside it. Adler saw in social life the basic fact, and not in the individual's socialization.

This ability for cooperation with others, the social interest or sociability *(Gemeinschaftsgefühl)*, is an innate potentiality. But in order to attain the high degree of cooperation necessary for survival, this innate potentiality must be developed far beyond its initial stage. Moreover, the desire for superiority must be coordinated with sociability and even subordinated to it.

Child Development and Style of Life

Adler strongly emphasized environmental influences but rejected the idea of an environmentalistic determinism. We are neither machines nor dictaphones. No experience, not even a traumatic one, is a cause of a future success or failure in one's life. Men give meaning to situations in accordance with their goals and life-styles. Happy or unhappy experiences influence one's life not by their happening but through the meaning one attaches to them. It is not the experiences that determine the individual's course of action but the conclusions he draws from them. His learning, remembering, and forgetting are selective, guided by the promise of success of his life style.

Adler compared heredity to a "supply of bricks" of various quality used by each individual in a different way in building his style of life. What matters most is not what one has inherited but *what one does* with what he has. Even art and genius are not underserved gifts of nature or inheritance but the individual's own creation.

All men strive toward superiority but each strives in a different way. No individual adjusts mechanically to his environment. The style of life is the expression of one's individuality, for each individual sees the goal of superiority in an individual, unique way.

The style of life controls the totality of the child's behavior. All aspects of life become coordinated and subordinated to this overall plan. The style of life represents the unity of the personality expressed in all facets of the individual's life: in his conscious and unconscious, in his thinking and feeling, acting and resting, in sex and in social interest.

In the first four or five years of life, the child absorbs the impact of his own body and environment. Then the creative activity or the style of life begins. In contradistinction to Freud, Adler assumed that the ego or the goal-directed creative power is formed after the first four or five years of life. Moreover, once the Adlerian ego or self or creative power of striving toward the goal is formed, the total personality is under its perfect control. In Adler's system, the ego represents the unity of personality.

Dreams and the Unconscious

Adler was influenced by the psychoanalytic theory of unconscious motivation, but his interpretation of the unconscious deviates considerably from Freud's concepts. The unconscious "is that which we have been unable to formulate in clear concepts. These concepts are not hiding away in some unconscious or subconscious recess of our minds, but they are those parts of our conscious, the significance of which we have not fully understood" (Adler 1935a: 10).

Conscious and unconscious are not opposites. Unconscious is whatever we don't fully understand. Unconscious represents the hidden instinctual drives and the not yet realized potentialities. Social relations form and modify the innate instinct, and the conscious (or consciousness) emerges out of this process of socialization.

According to Adler, all dreams reflect a problem. The subject of a dream is always a problem, the solution of which the individual, in his waking life, is not sufficiently prepared for, is hesitant about, and does not know how to solve with logic and common sense in accordance with the demands of the community. He therefore cannot solve it in his dream either, except by the abolition of logic and common sense and the neglect of social demands and necessity. Every dream, then, poses one or more alternative solutions that are devoid of social feeling, logic, and common sense. They are, therefore, *asocial* solutions; they indicate the style of life of the individual as well as the particular life problems he fears. This holds for all dreams of patients; it is equally true for the dreams of most other individuals. Adler said that a rare person may have such a high degree of social feeling and be so integrated that in the dream he may continue his *social* search for a solution and even come with a *social* answer; this, however, is rare.

The purpose of the use of symbols and metaphors in the dream is to conceal the real meaning of the dream from the dreamer. Were he to understand that the solution of his problem, so easily accomplished in the dream and thereby erasing his feeling of inadequacy is, in actuality, a self-deception, then the purpose of the dream would be nullified.

Dreams of falling usually indicate a fear of loss of prestige or position and a warning to oneself not to climb too high. Dreams of flying indicate an active, ambitious person and an attitude that difficulties can easily be overcome. Dreams of paralysis indicate that he considers the problem to be insoluble. Dreams of taking examinations, if accompanied by anxiety, mean that he feels unprepared for the task ahead; if without anxiety, it may mean self-encouragement: "I did it before—I can do it again." Dreams of dead people often mean that he hasn't buried his dead yet and still remains under their influence; occasionally they may indicate thoughts of joining the dead, and thus imply suicidal tendencies. Dreams about being improperly clothed indicate fear of being found out in an imperfection which one believes one has to hide. If a dreamer pictures himself as a mere spectator in his dream, it is safe to assume that he will be satisfied with an onlooker's role in waking life, too. Sexual dreams may have many different meanings depending on their content, context, and feelings in them. Some may indicate unpreparedness for sexual relations and a warning against them, others may be a training for them. Dreams of sex with others than one's partner will indicate hostility toward the partner. Homosexual dreams may be a training against relationships with the other sex or a self-reassurance that one can do without the other sex, (K. Adler 1967: 314–316).

Personality Structure

Adler used the terms character and personality interchangeably. One's personality structure is determined by the two driving forces of self-assertion *(Geltungsfrieb)* and sociability *(Gemeinschaftsgefühl)*, which, combined, form one's style of life. As mentioned before, Adler rejected the notion of determinism and ascribed to the individual the creative ability of setting his own goals and shaping his own life. The attitude toward existing social values, which resembles Edward Spranger's psychology, was described by Adler under the name "degree of activity."

Every individual sets up a fictional ideal of himself. "The fictional, abstract ideal is the point of origin for the formation and differentiation of the given psychological resources into preparatory attitudes, readinesses, and character traits. The individual then wears the character traits demanded by his fictional goal, just as the character mask (persona) of the ancient actor had to fit the finale of the tragedy." Moreover, the ego or self or creative force guides the individual toward the ideal of himself. "In every case the point of the self-ideal *(Persönlichkeitsideal)* posited beyond reality remains effective. This is evidenced by the direction of the attention, of the interests, and of the tendencies, all of which select according to points of view given in advance" (Adler 1956: 94).

Adler's *self-ideal* corresponds to Freud's ego-ideal and later superego, but it plays a less important role in Adler's personality structure as compared to Freud's fully developed personality model. The way the self-ideal was worded by Adler indicated his unwillingness to consider this "ideal" as a part of the mental apparatus. It rather indicates a striving than an accomplishment.

In the striving toward the goal, character traits and complexes are formed. "The character traits are . . . the outer forms of the movement line of a person. As such they convey to us an understanding of his attitude toward the environment, his fellow man, the community at large, and his life problems. They are phenomena which represent means for achieving self-assertion. They are devices which join to form a method of living" (p. 219).

Human emotions are intensifications or accentuations of character traits; they are forms of psychological movement, limited in time. They are purposeful and they serve the purpose of improving the situation of an individual. Joy, anger, fear, and sorrow lead to an intensified movement of the individual, who makes an effort to come closer to his goal.

Typology

Adler reinterpreted Hippocrates' theory of the four temperaments, the sanguine, choleric, melancholic, and phlegmatic. The sanguine type is the most healthy, believed Adler. Since the sanguine person was not subjected to severe deprivations and humiliations, he has little if any inferiority feeling. He is capable of striving toward superiority in a happy and friendly manner.

The choleric is tense and aggressive. His striving for superiority and power involves a great expenditure of energy. He goes about attaining his goal in a direct, aggressive manner. His social adjustment is rather poor.

The melancholic type succumbs to his inferiority feeling to such an extent that he lacks initiative in overcoming his obstacles. He is worrisome and undecided and lacks self-confidence and courage to take risks. He is not antisocial but does not participate too much in social interaction. The phlegmatic type is depressed, slow, sluggish, not impressed by anything, and unable to make an effort for improvement.

In 1935, two years before his death, Adler introduced a new typological system based on the concepts of degree of activity and the social interest as the two guiding principles in the grouping of personality types. The first type displays a great deal of activity in pursuing his goal but lacks social interest. His lack of consideration for others makes him act in an antisocial manner. He is the *dominant* or *ruling* type. This type corresponds to Hippocrates' choleric type. The second type is the *getting* type. He lacks both activity and social interest and expects others to take care of him. He himself lacks initiative and consideration for others. The getting type resembles the phlegmatic type. The third type is the *avoiding* one. Instead of struggling for success and superiority, he stands still, undecided. Both his activity and social interest are very small. The avoiding type resembles the melancholic. The fourth type is the *socially useful* one. He is active but his activity is in harmony with the needs of others and beneficial for them. The useful type resembles the sanguinic (Adler 1935).

Psychopathology

The parent-child relationships are, ultimately, the single most important factor in the development of mental disorder. All children try to test all sorts of methods to gain or regain their sense of security, mastery, and superiority, such as: expressions of fear to get support and affection (anxiety); displays of crying or pouting to force others to do their bidding and to feel sorry for them (depression); attempts at hurting thmselves in order to blame, attack, or accuse others (suicide); construction of magical thought patterns or magical orders to undo feelings of impotence or insecurity and to avoid having to deal with the real problem (compulsion); denial of the reality of an occurrence and its unbearable feelings (schizoid); accusations of others to escape responsibility (paranoid); flight into fantasy to attempt life in a better, private world instead of the real one that is too difficult (psychoses).

Adler described three categories of children prone to mental disorders:

1. Children who are born with organic deficiencies; those with motor, digestive, respiratory weaknesses, with visual or auditory handicaps, clumsy, weak or sickly children, etc. For any small adaptation to life, for the smallest satisfaction, they have to make such strenuous and concentrated efforts that their attention, constantly on themselves and their effort for adaptation and survival, have little left for concentration of their interest on others. (When you are drowning in a lake, you are not able to appreciate the beautiful scenery around you.) Others are of interest to them only for whatever help they can get from them, in their

frantic efforts. Dependent and exploitative relations are frequently the result, unless a very good and secure relationship has nevertheless enabled them to develop sufficient interest in other people.

2. Children who develop what Adler called a "pampered style of life." The choice of the word "pampered" may perhaps be unfortunate; first, because it denotes coddling and loving, which are generally necessary in the upbringing of children; and second, because it may imply that there was a stupid evil parent who committed the "great sin" of pampering the child. . . . They look for easy triumphs, praise, and admiration; they never trust their own powers to achieve a desired victory, are highly envious of others and prone to depreciate them, including those on whom they depend for help. As children they already show the precursors of such neurotic reactions as anxiety, depression, compulsion, etc.; they have various asocial habits and are always a burden instead of a help in the family. Actual overprotection and pampering may be a strong incentive for a child to make such interpretations and to develop a pampered style of life. The youngest child, for instance, the baby in the family, may easily develop such ideas about life when parents and other children do everything for him. (Sometimes, however, a child, in reaction, may rebel against such pampering.) In the first category mentioned, the physically handicapped children may be additionally endangered by a quite understandable overprotection given them by the parents. Also, the existence of impressively stronger siblings and their outstanding successes may easily impress a child with the impossibility of ever equalling them, and thus lead them to the pursuit of devious ways in competition with them, such as: seductiveness, pleas for sympathy by stressing their weaknesses, arousing guilt in others, or constructing alibis and excuses for their failure, and shifting the responsibility for it onto others. . . .

3. Children who have been neglected, rejected, or hated—the unwanted children. They, too, are prone to believe very early in life that the only ideal to strive for is to get support and service from others; but having been devoid of any experience with love, friendship, and cooperation, they not only do not know how to use these methods of social relations, but are, in fact, highly suspicious of them. (Adler 1967: 308–309)

BIBLIOGRAPHY

Adler, A. 1907. *Study of Organ Inferiority and Its Psychical Compensation.* S. E. Jeliffe, tr. New York: Nervous and Mental Disease Publishing Co., 1917.

——1912. *The Neurotic Constitution.* Introduction by W. A. White. B. Glueck and J. E. Lind, tr. New York: Moffat, Yard, 1917. Plainview, N.Y.: Books for Libraries, 1974.

——1929. *Problems of Neurosis: A Book of Case Histories.* Philip Mairet, ed. New York: Harper and Row Torchbooks, 1964.

——1956. *The Individual Psychology of Alfred Adler: A Systematic Presentation in Selections from His Writings.* H. L. and Rowena R. Ansbacher, eds. New York: Basic Books.

——1964. *Superiority and Social Interest: A Collection of Later Writings.* H. L. and Rowena R. Ansbacher, eds. New York: Viking Press, 1973.

Adler, K. A. 1967. Adler's individual psychology. In B. B. Wolman, ed., *Psychoanalytic Techniques,* pp. 299–337. New York: Basic Books.

Ansbacher, H. L. 1967. Life style: a historical and systematic review. *Journal of Individual Psychology,* 23: 191–212.

——1968. The concept of social interest. *Journal of Individual Psychology,* 24: 131–149.

Ansbacher, H. L. and R. Ansbacher, eds. 1956. *Individual Psychology of Alfred Adler.* New York: Basic Books.

Orglen, H. 1963. *Alfred Adler: The Man and His Work.* New York: New American Library.

Schulman, B. H. 1973. *Contributions to Individual Psychology.* Chicago: Alfred Adler Institute of Chicago.

Vaihinger, H. 1925. *The Philosophy of "As If"; A System of Theoretical, Practical, and Religious Fictions of Mankind.* New York: Harcourt, Brace.

[FIFTEEN]
Jung's Analytic Psychology

Carl Gustav Jung, the first president of the International Psychoanalytic Association, broke away from Freud in 1914. Initially Jung rejected Freud's theory of libido and sexuality, and as years went by, Jung developed a new psychological system vastly different from Freud's system, leaving perhaps a single link between the two, namely the idea of unconscious motivation.

On this issue Jung proved to be *plus catholique que le pape*. Jung was more concened by, more deeply involved in, and more enchanted by the primordial beauty of the unconscious. As his work progressed, he attached more and more importance to the unconscious.

Freud viewed human nature in the perspective of unconscious, preconscious, and conscious, the conscious being the controlling force. Jung definitely rejected the rule of the conscious.

Freud persistently applied scientific analysis to all "mental provinces," including the unconscious. A great deal of his work was devoted to the primary processes, but the method he used was always the rational method and empirical observation. It was always the conscious part of Freud's mind that analyzed dreams.

Jung was deeply involved with the unconscious phenomena and bluntly refused to apply to them the method of scientific analysis. He insisted that we must make sure that "we do not foist conscious psychology upon the unconscious," for the unconscious cannot be represented in terms of thinking, reasoning, etc. Jung refused to accept the idea that the unconscious processes could be perceived by and represented in terms of the conscious mind.

In a sense, Jung violated the Freudian principle of reality testing by saying that the unconscious "affects us just as we affect it. In this sense the world of the unconscious is commensurate with the world of outer experience" (Jung 1928a: 198).

Freud was a determinist, Adler a purposivist. For a long time Jung seems to have accepted both principles. Human mind, he used to say, is guided by both, for it is, on the one hand, a "precipitate of the past," and, on the other hand, the psyche "creates its own future."

Jung's discussion of this problem was devoted less to the search for truth and more to the analysis of the possible implications of causation and teleology for human attitudes to life. Jung was not sure that either of them could be proved or disproved; he accepted somewhat the Kantian position and believed that both principles are methods of cognition rather than laws of nature. The acceptance of causation would however, make men feel unhappy, pessimistic, and unable to undo their past, while purposivism leaves more freedom to the individual. Consequently, decided Jung, man lives by aims as well as by causes. Jung did not see any contradiction in these two principles. Moreover, while he was opposed to the deterministic causalism, he introduced a fatalistic predestination theory through the eternal factors of ancestral history, which influence the actions of each individual.

Mental Energy

Jung accepted the idea that mental energy is a continuation of physical energy and that each of the two types of energy could be transformed into the other.

> Life takes place through the fact that it makes use of natural physical and chemical conditions as a means to its existence. The living body is a machine that converts the amounts of energy taken up into its equivalents in other dynamic manifestations. One cannot say the physical energy is converted into life, but only that the transformation is the expression of life. . . . All the means employed by an animal for the safeguarding and furthering of his existence, not to speak of the direct nourishment of his body, can be regarded as machines that make use of natural potential in order to produce work. . . . Similarly, human energy, as a natural product of differentiation, is a machine; first of all a technical one that uses natural conditions for the transformation of physical and chemical energy, but also a mental machine using mental conditions for the transformation of libido. (Jung 1928b: 46)

Mental energy or libido is used basically for self-preservation and preservation of the species. When these biological needs are satisfied, it is available for the pursuit of other goals, such as cultural, social, or creative needs of the individual.

Jung maintained that human behavior is dominated neither by the all-powerful sexual libido of Freud nor by the mastery drive of Adler. There is only "undifferentiated life energy," which expresses itself at one time in the pursuit of sensual pleasure and at another time in the striving for superiority, artistic creation, play, and other activities.

Dialectics

One of the main principles of Jung's theory is the *principle of opposites*, or dialectics. Life is construction and destruction, creation and decay, waking and sleeping. It is a cosmic principle, reminiscent of the tenets of the ancient Greek philosopher Heraclitus and the idealistic German philosopher Hegel. It is the dialectic law of development through opposites, through the swinging from one extreme to the other, from a thesis to its antithesis. After oscillations comes stability; the deeper was the conflict, the more profound the stability.

This law serves as the guiding principle of the human mind:

> Everything human is relative, because everything depends on a condition of inner antithesis; for everything subsists as a phenomenon of energy. Energy depends necessarily on a pre-existing antithesis, without which there could be no energy. There must always be present height and depth, heat and cold, etc., in order that the process of equalization—which is energy—can take place. All life is energy, and therefore depends on forces held in opposition. (Jung 1928a: 78)

Entropy

The greater the conflict between the opposites, the more mental energy comes out of them. Discharge of energy is caused by the state of inner conflict within a given system. The distribution of mental energy follows the principles of equivalence and entropy. The principle of conservatism of energy or *equivalence* states that energy removed from one area will appear in another. Jung's theory is a faithful exposition of this principle. His principle of *entropy* is borrowed from the second law of thermodynamics. If one part of the personality is charged with a heavy load of libido and another with a low load, libido will move from the former toward the latter.

The two main rules of libido movements are *progression* and *regression*. When the opposite forces within the system are balanced and the psyche is in a state of equilibrium, the libido moves smoothly, from the unconscious layers of personality toward the conscious in the process of progression. "During the progression of the libido the pairs of opposites are united in the coordinated flow of psychical processes. Their working together makes possible the balanced regularity of these processes, which, without this reciprocal action, would be one-sided and unbalanced". The individual whose libido is in state of progression experiences the feeling of pleasure, happiness, and well-being. Jung calls it the "vital feeling."

In some situations the progression of the libido is thwarted. Dammed up, its flow is stopped. The opposites united in the flowing libido fallapart. The inner conflict mounts, great amounts of energy are generated, and "the vital feeling that was present before disappears and in its place the psychic value of certain conscious contents increases in an unpleasant way; subjective contents and reactions press to the fore and the situation becomes full of affect and favorable for explosions" (Jung 1928b: 46).

The Conscious and the Unconscious

The conscious plays in Jung's theory a secondary role as compared to the unconscious. The conscious is useful in adjustment to the environment. "The essence of consciousness is the process of adaptation which takes place in the most minute details. On the other hand, the unconscious is generally diffused, which not only binds the individuals among themselves to the race but also united them backwards with the peoples of the past . . . and is . . . the object of a true psychology" (Jung 1916: 199).

The psyche is composed of three parts, the conscious, the personal unconscious, and the collective unconscious. The conscious and the unconscious are opposites that balance each other in a *reciprocal relativity*. The tension between the conscious and unconscious parts of the mind sets free psychic energy.

The personal unconscious is the superficial layer of the unconscious. It encompasses all the forgotten memories, subliminal perceptions, and suppressed experiences. In addition, it contains "fantasies (including dreams) of a personal character, which go back unquestionably to personal experience, things forgotten or repressed" (Jung 1949: 102).

Dreams

According to Jung, dreams are undisguised manifestations of unconscious creative activity. In addition, dreams perform compensatory functions. "When in the course of analysis, the discussion of conscious material comes to an end, previously unconscious potentialities begin to become activated and these may easily be productive of dreams" (Jung 1954: 100). Entropy, the law of compensation, is one of the basic laws in Jung's theory, and the task of dreams is to compensate for conscious attitudes.

Jung maintained that dreams contain more than wish fulfillment and they reveal several aspects of one's personality. "There are, it is true, dreams which manifestly represent wishes or fears, but what about all other things? Dreams may contain ineluctable truths, philosophical pronouncements, illusions, wild fantasies, memories, plans, anticipations, irrational experiences, even telepathic visions," wrote Jung (1966: 147).

Jung reported a dream told to him by one of his older colleagues who frequently expressed doubts about Jung's method of dream interpretation.

> I am climbing a high mountain, over steep snow-covered slopes. I climb higher and it is marvelous weather. The higher I climb, the better I feel. I think, "If only I could go on climbing like this forever!" When I reach the summit my happiness and elation are so great that I feel I could mount right into space. And I discover that I can actually do so: I mount upwards on empty air and awake in sheer ecstasy. (p. 151)

Jung warned his friend not to go mountain climbing unless accompanied by two guides, but the old man did not heed Jung's warnings. Two months later the man went in the mountains, was buried by an avalanche, but luckily was saved by an Alpine patrol. Three months later, he climbed mountains with a friend and lost his balance. A mountain guide saw him from afar and saw how he fell on the head of his friend who was standing below. Both were crushed to death. Jung admitted that this dream could be interpreted as a wish to commit suicide or homicide or both. But it also represented an uncanny prediction of the future.

According to Jung, there is little reason to assume that the manifest dream is merely a façade and the latent dream content is the essence. Jung believed that the whole meaning of a dream is represented in the manifest dream.

> What Freud calls the dream-façade is the dream's obscurity and this is really only a projection of our own lack of understanding. . . . The dream is a statement, uninfluenced by consciousness, expressing the dreamer's inner truth and reality "as it really is" not as I conjecture it to be and not as he would like to to be—but "as it is." (p. 142)

The Jungian technique of dream interpretation was comprised of the following steps: analysis of the present situation of consciousness, description of preceding events, interpretation of the subjective context and archaic motives in the dream, and comparison with objective data obtained from third persons.

Symbolism

Jung's system introduced a great many concepts related to symbolism. He drew a sharp line between sign and a symbol. A sign represents a known object or action, while a symbol is a description of an unknown fact (1923: 601).

Symbols originate in the unconscious. The energy for the formation of symbols is derived from the excess of libidinal energy. The satisfaction of instinctual demands is the main channel for the libido, but, as the history of civilization has proven, "man possesses a relative surplus of energy that is capable of application apart from the natural flow. The fact that the symbol makes this deflection possible proves that not all the libido is bound up in a form that enforces the nature flow, but that a certain amount of energy remains over, which could be called excess libido" (Jung 1960: 47).

Moreover, while symbols are created by excess of energy, they represent a higher and more effective level of human activities.

Jung distinguished two types of symbols, as follows:

> The former [natural symbols] are derived from the unconscious content of the psyche, and they therefore represent an enormous number of variations on the essential archetypal images. In many cases they can still be traced back to their archaic roots—i.e., to ideas and images that we meet in the most ancient records and in primitive societies. The cultural symbols, on the other hand, are those that have been used to express "eternal truths," and that are still used in many religions. They have gone through many transformations and even a long process of more or less conscious development, and have thus become collective images accepted by civilized societies. (1964: 93)

Apparently, in Jung's system the symbols are elements of complex and unconscious facts not yet clearly understood by consciousness.

Since the symbols originate in the unconscious, "it is not possible to discuss the problem of symbol formation without reference to the instinctual processes, because it is from them that the symbol derives its motive power. It has no meaning whatever unless it strives against the resistance of an instinct" (Jung 1956: 228).

The unconscious supplies, as it were, the raw material for symbols, but it is the conscious that creates the symbols; it is only in the interaction between the two that we can really understand the symbol. Therefore, "inasmuch as the symbol proceeds from the highest and latest mental achievement and must also include the deepest roots of his being, it cannot be a one-sided product of the most highly differentiated mental functions, but must at least have an equal source in the lowest and most primitive notions of his psyche. . . . The symbol has one side that accords with reason; but it also has another side that is inaccessible to reason; for not only the data of reason, but also the data of pure inner and outer perception, have entered into its nature" (Jung 1923: 607). The fact that the conscious and the unconscious together shape the symbols represents the main function of the symbol, namely the *reconciliation of opposites*.

Such a dialectic *reconciliation of opposites* is one of the core concepts in Jung's system. The symbol which serves the combination of thesis and antithesis is called "living symbol," and the process of reconciliation and merger is called "transcendent function."

Being created by both the conscious and the unconscious, symbols are both real and unreal. Were a symbol only real, it would have been a real phenomenon, "and therefore removed from the nature of a symbol. Only that can be symbolical which embraces both. If altogether unreal, it would be mere empty imagining, which, being related to nothing real, would be no symbol" (Jung 1923: 141).

The unconscious expresses its experiences in symbol, but it is up to the consciousness to accept this expression. Jung believes that in both the lives of individuals and the entire Western civilization there is a tendency to ignore the voice of the unconscious. The gulf between the conscious and unconscious aspects of life is likely to make the unconscious destructive and dangerous.

Since the source of the symbol is attributed to unconscious processes, according to Jung, it would seem that the formation of the symbol requires a regression into the unconscious. Jung welcomed such a regression provided it leads to a "transcendent function" which reconciles the dissociative elements of the conscious and unconscious, thus facilitating the developmental process and leading to the formation of the unified self. The symbols mediate between the conscious and the archetypes of the collective unconscious, thus becoming the basis for all of civilization. Symbols play the unifying role in several areas of human behavior and foster the development of individuals and culture as a whole.

Dreams use "natural" symbols. The symbols may represent different things in different individuals, and even in the same individual at different times. All dreams and their symbols originate in the collective unconscious, they themselves from the basis for collective or cultural symbols. "Judging from the close relation of the mythological symbol to the dream symbol . . . it is more than probable that the greater part of the historical symbols arise directly from dreams" (Jung 1928: 54).

Complexes

The psyche, according to Jung, is comprised of several discrete systems which interact with each other, among them "the ego"; "the personal unconscious," with its associated "complexes"; "the collective unconscious," with its "archetypes"; "the persona;" "the shadow"; and the "anima" and "animus."

The ego is the conscious mind. It is composed of various subjective qualities including memories, thoughts, perceptions, and feelings, and it is responsible for a person's feeling of identity, or sense of uniqueness from others. The conflict between the conscious and the unconscious parts of the psyche generates psychic energy; this energy cannot be set under the control of the consciousness, and units so loaded with mental energy thus start their independent life.

These units of energy, set free, attract ideational content and form independent entities or constellations or *complexes*. Complexes are *autonomous partial systems* in the human mind; they form split personalities within a given personality; they indicate the relatively independent inclinations within the same individual.

In dreams these separate inclinations appear in a disguise of characters. It is as if the dreamer were experiencing the various characters that exist in his mind beyond his control. These "psychic splits" are universal; each individual is composed of several relatively independent individualities. Only when these complexes get too far apart from each other and from the rest of the psyche is mental health jeopardized. When a complex reactivates elements of the deeper layers of the unconscious, the archetypes, and takes over control of the personality, a psychosis has developed.

One of these complexes is *persona*. *Persona* is Latin for "mask." Actors had to wear different masks for different plays; analogously, said Jung, each individual wears different masks playing different roles in relationship to different people. The same man is a son to his father, a husband to his wife, an employer to his employees, a father to his children. In each situation he puts on a different mask. He can never be the totality of all his conscious and unconscious forces, and in each situation another part of his personality comes to the fore. The various relationships elicit responses, and the degree of consistency in the different masks of one person depends on the inner integration.

The persona represents the conscious attitudes of the individual toward the outer world. It must be, therefore, related to the ego. But the *ego* in Jung's theory is merely a certain "condition of consciousness." It is a "complex of representations which constitutes the centrum of any field of consciousness and appears to possess a very high degree of continuity and identity." Accordingly the ego forms the kernel of one's *persona,* which represents the attitudes of the individual toward the outer world. The strong, domineering qualities in the individual gather together into the conscious ego and the ego is drawn into the persona. In some cases the entire "conscious area as seen by others, i.e., the persona, becomes identical with the ego. This is the case with individuals whose entire life is guided by a certain focal idea or talent" (Jung 1923: 564).

The weak and least adapted tendencies also gather together to form an unconscious complex, the *shadow.* The shadow contains the urges and wishes which cannot be approved of by the conscious ego. It is a personality within a personality. The shadow represents the forbidden sexual and aggressive impulses; it forces the individual to irresponsible and dangerous actions; it embarrasses the ego by tactless and stupid blunders; it gives the individual unpleasant, often weird feelings. The shadow has its own psychic energy. If strong enough, it may pierce the conscious and take over control. In such a case, mental disorder develops.

One of the main tendencies of the shadow is projection. "We still attribute to the 'other fellow' all the evil and inferior qualities that we do not like to recognize in ourselves. That is why we have to criticize and attack him" (Jung 1933: 163).

The idea of the mother forms the core of the mother complex. This complex is a result of one's personal experience with his mother and his inherited legacy pertaining to the experiences that his ancestors have had with their mothers, thus the mother complex borders on the collective unconscious. The nucleus of the complex becomes associated with feelings, cognitions, and memory traces related to mother. The greater the amount of energy possessed by the nucleus, the greater the number of experiences that it will engulf.

Archetypes

Jung believed that acquired traits and cultural patterns are transmitted by heredity. In each individual, he wrote, there are hidden "the great 'primordial images,' those potentialities of human representations of things, as they have always been, inherited through the brain structures from one generation to the next." Jung did not deny that besides these collective deposits, which contain nothing specifically individual, the psyche may also inherit "memory acquisitions of a definite individual stamp."

These primordial images are called "archetypes." Jung was much influenced by the French social scientists Emile Durkheim and Lucien Lévy-Bruhl and was especially impressed by the idea of l'esprit du corps ("collective spirit"). This collective spirit must be unconscious and manifests itself through the individual mind.

Archetypes reveal themselves in tribal lore, but the way they are expressed changes as they are elaborated in the esoteric transmission of traditional rites. The archetypal images are present in folkways, customs, and rites, reflecting the people's past and their ways of existence. The traditional beliefs, legends, and myths carry in an individual's psyche the memories of his ancestors. These memories are much alive in the individual's collective unconscious. Primitive men assimilated all outer sense experience to inner, psychic events and their myths are primarily "psychic phenomena that reveal the nature of the soul."

The contents of the archetype nature are manifestations of the collective unconscious. "They do not refer to anything that is or has been conscious, but to something *essentially unconscious*. In the last analysis, therefore, it is *impossible* to say *what they refer to*. . . . The ultimate core of meaning may be circumscribed but not described. Even so, the bare circumscription denotes an essential step forward in our knowledge of the pre-conscious structure of the psyche, which was already in existence when there was as yet no unity of personality" (Jung and Kerenyi 1949: 104).

The archetypes represent the memories of a race and are deep unconscious forces in the deep layers of the mentality of the members of a race. Since Jung believed in the heredity of acquired traits and culture patterns, he ascribed to certain races inherited patterns. In German patients he observed archetypes stemming from the pre-Christian era; the symbol of the ancient German god Wotan, which expresses violence and cruelty, was found in the unconscious of German people.

The archetypes form the core of autonomous partial systems, independent of consciousness. Once an archetype is stirred up, it develops into an autonomous partial system and takes "possession" of an individual;

the individual is mentally sick. These autonomous systems are *demons* and overwhelm the mind with their irresistible power.

In normal cases these systems are expressed by symbols. Symbols can move up to the conscious, carrying the elements of the archetypes and the collective unconscious, as expressed in mythology. Mythology contains the dreams of mankind. Jung called the collective symbols *motifs* and believed that the motifs were the links between the individual and the cosmos.

Child Hero

One of the archetypes is the child hero. The child Moses is saved out of the waters of the Nile; the child Christ was hidden by Mary and Joseph. The child symbolizes the emergence of the self; the libido regresses into the unconscious, enters the womb (in a symbolic analogy to sex), and emerges again in childbirth.

> "The 'child' is born out of the womb of the unconscious, begotten out of the depths of human nature, or rather out of Living Nature herself. It is a personification of vital forces quite outside the limited range of our conscious mind; . . . a wholeness which embraces the very depth of Nature. It represents the strongest, most ineluctable urge in every human being, namely the urge to realize itself. (Jung and Kerenyi 1949: 123)

This archetype in a positive form of self-fulfillment, satisfaction of conscious goals, and attainment of the futurity potential, or its negative form of abandonment and insignificance of human life, provides a fundamental framework for the understanding of the human mind. Jung believed that the child archetype is a collective symbol of the self which manifests itself in development of the total personality. For instance, the Child Jesus directs human faith through the miracle, mystery, and authority now manifest in the Christian religion. The historical fact of Christ's human birth is psychologically insignificant, but the divine nature of Christ and his image of the "Child God" carries the archetypical significance.

The human concept of oneself is derived from his childhood experiences, and the image of one's childhood is of crucial importance in the formation of one's self. The child archetype represents in an individual's mind the anticipation of future developments, and especially the ultimate synthesis of conscious and unconscious elements in the personality. The child archetype unites the opposites and serves as a mediator, bringer of healing, and makes a whole out of the two elements.

An archetype may have, however, a positive and a negative meaning. On the negative side, the child archetype represents the idea of insignificance of human life, exposure, abandonment, or danger. The child archetype is also an image of powerlessness and helplessness of the life urge. The child archetype then becomes symbolic of the paradoxes of life. The paradoxes of human life, the polarity of good and evil, the unconscious threats coming from the instincts that jeopardize one's own, are incorporated in the archetype of child as a human being which lives in the future and the past simultaneously. The magical elements of the child archetype reflect the fact that the human conscious is unable to reach beyond these opposites, and therefore attributes an aura of magic to this archetype, which unites the opposites. The child archetype represents development toward independence, and independence implies separation of oneself from its origins. Small wonder that the child archetype carries both the positive aspect of self-fulfillment and realization and the negative aspects of abandonment, danger, and pain.

Water

Water is another archetype with a positive and negative form. The archetype of water represents the relation between the individual or personal unconscious and the universal or collective unconscious. Water is "the commonest symbol of the unconscious. The lake in the valley is the unconscious, which lives as it were, underneath consciousness" (Jung 1959: 6). Water also stands for a vital substance for life, as the fluid which courses through the body. One's survival depends on maintaining this fluid in the body, and all human beings are motivated by an unconscious drive for water, the fundamental bodily need.

Animus and Anima

The archetypes animus and anima merit special attention for they represent masculinity and femininity respectively. According to Jung's idea of reciprocal relativity, a person who is consciously male carries in his unconscious feminine elements and vice versa. This principle applies to both collective and personal unconscious. Thus, if in one personal unconscious the persona complex stands for masculine traits, the shadow contains feminine ingredients.

A man is both masculine and feminine, and even in the most masculine man, there are soft feminine emotions. Man tends to repress his feminine traits and they are hidden in his unconscious mind. When a man looks for a woman to love, he usually chooses the one who resembles his *own* unconscious feminine traits, and that is why men quickly identify the women that they are looking for.

The persona complex plays a compensatory role in regard to the anima. In order to compensate for the unconscious, tender feminine feelings of the anima, men tend to underscore the identification with their masculine persona.

Anima. Anima means soul, spirit, the essence of life. The archetype anima is the feminine element in man and creator of all life. The anima is personified in mythology by mermaids, sirens, and wood nymphs. The anima has positive and negative sides; it is at the same time good and evil, something to be both feared and cherished. Anima is "the urge to life."

> Being that has soul is living being. Soul (anima) is the living thing in man, that which lives of itself and causes life. Therefore God breathed into Adam a living breath, that he might live. With her cunning play of illusions the soul lures into life the inertness of matter that does not want to live. She makes us believe incredible things, that life may be lived. She is full of snares and traps, in order that man should fall, should reach the earth, entangle himself there, and stay caught, so that life sould be lived; as Eve in the Garden of Eden could not rest content until she had convinced Adam of the goodness of the forbidden apple. (Jung 1968)

The anima is the spontaneous element in psychic life and encompasses man's drives and impulses. The anima is the driving force in life. It is unconscious but it give rise to consciousness. Anima is like life itself and contains both the good and the evil.

The male infant, being born with his feminine archetype, gets his first experience of a woman through his mother. This experience is, however, entirely subjective and does not depend on how the mother acts, but how her actions are perceived by the child. The mother's image is shaped by the child's *inherited* capacity to produce an image of a woman—the archetype anima.

The image of woman exists in man's collective unconscious and enables him to apprehend the nature of a woman. Men tend to project the inherited image of anima on all women he meets and attributes his own unconscious anima feelings to women. According to Jung, the anima image lends the mother a super human glamour in the eyes of the son, but gradually becomes tarnished by reality and sinks back into the unconscious. However, the anima image does not lose its vigor and is ready to project itself at the first opportunity, whenever a woman makes a strong impression on the man. The unconscious anima can make a man soft or jealous, moody or ambitious and vain. The projection of the anima on a man's wife can make him childish, dependent, and subservient, or on the other hand, tyrannical, truculent, morbidly preoccupied with his masculinity. In the projection, the wife becomes the guardian of his unconscious and assumes the role of his mother. The man who projects the mother's image on his wife becomes submissive, as if seeking his mother's protection. His fear of his own unconscious femininity forces him to accept this subservient position.

The anima, like all other elements in Jung's system, has two sides. It represents on the one hand the pure, the good, the noble goddesslike figure, but on the other side, the anima is the archetype of a prostitute, a seductress, or a witch, and she guides the man towards the spiritual heights and the depravities of life. The direction the anima leads to depends on the man, and his choice of the modes of instinctual behavior symbolized by the anima image.

Animus. The animus represents the masculine element in woman. The animus is derived from the collective image of man that woman inherits, the woman's experiences with men and the latent masculine element in woman. The anima produces emotions and moods while the animus creates opinions. According to Jung, the animus resembles an assembly of dignitaries who lay down allegedly "rational" judgments. On closer examination these judgments turn out to be largely sayings and opinions transmitted unconsciously from childhood and misrepresented as canon of truth, justice, and reasonableness.

The animus of a pretty woman makes her perceive the man as fatherly and professorial. Intellectual women stress highbrowism, which, according to Jung, is simply harping on a rather irrelevant point. Un-attractive women are at the mercy of their animus, and desperately try to infuriate men. Women who are at the mercy of the animus may lose their feminine persona.

Women are primarily concerned with their husbands, children, and friends. Men are mostly concerned with business and politics. To a woman, her family represents her entire life, while to a man, the family life is but the means to an end. A woman thinks of her husband as the only man in the world, but a man never sees in his wife the only woman.

The animus can be personified by any male figure, depending upon the woman's developmental stage. Also the animus has a positive and negative side. On one hand, the animus can stir a woman to search for knowledge and truth, and guide her onto a productive and useful way of life. On the other hand, the animus can elicit aggressive impulses in women and make them crave for power and control. The Egyptian queen Cleopatra is a case in point. Cleopatra ruled in an absolute and authoritarian manner; certainly she was possessed by the animus.

As Jung went on formulating and changing his theory, he attached more and more importance to the anima. Initially Jung viewed anima as one of the complexes, but later made it stand as a symbol of the entire unconscious.

The Mother Archetype

The mother archetype represents solicitude and sympathy and the magic power of the female. All that is helpful and benign, all that cherishes and sustains and that fosters growth and fertility is incorporated in the mother archetype. Like all other archetypes, the mother archetype is also ambivalent. The mother archetype also represents the hidden and dark aspects of the mind. The negative mother archetype is symbolized by the *praying mantis*, the insect that devours her mate as he fertilizes her for the birth of her offspring. The mother archetype symbolizes seduction and greed on the one side, compassion and charitable attitudes on the other. The mother archetype is of great importance in personality development because the child's first contacts with the world are with his mother.

The Ego and the Self

Whenever the individual is capable of raising the unconscious, powerful, libido-loaded, miraculous, *mana*-like anima from the unconscious into the conscious, the total personality undergoes deep and profound changes. The anima, elevated from unconscious into conscious, loses its tremendous load of libido. It becomes depotentiated and loses the power of *possession*.

The mental energy releaed by the now-conscious anima is free, unbound. It is a tremendous quantity of energy which is neither conscious nor unconscious. This libido becomes the central point of personality between the conscious and unconscious and is called *self*. The self, said Jung in his *Two Essays*, "has come into being only very gradually and has become a part of our experience at the cost of great effort. Thus the self is also the goal of life" (Jung 1928: 268).

The conflicting elements of conscious and unconscious come to harmony in the self. This integration of personality through the emergence of self is found in the art and religion of several people, especially of the Far East. It is represented by the *mandala* symbol, in the form of a square or a circle with a central point. The mandala, according to Jung, represents the reconciliation of opposites, the fusion of the conscious and the unconscious.

The various complexes and systems within the personality strive toward expansion and expression. If one of them attracts too much libido, the other centers will oppose it and stir inner conflict. In normal individuals proper outlets are provided for the various aspects of personality. The process by which development of the respective parts of the personality is facilitated is called *individuation.* The aim of individuation is "to free the self from the false wrappings of the Persona on one hand, and from the suggestive power of the unconscious image (i.e., the anima) on the other" (Jung 1928: 185).

This diversifying individuation is a step forward toward integration of the personality and development of the *self.* The *transcendent function* takes place by which the *already differentiated systems unite into one whole.*

> If we picture the conscious mind with the ego as its center, as being opposed to the unconscious, and if we now add to our mental picture the process of assimilating the unconscious, we can think of this assimilation as a kind of approximation of conscious and unconscious, where the center of the total personality no longer coincides with the ego, but with a point midway between the conscious and unconscious. This would be the point of a new equilibrium, a new centering of the total personality, a virtual center which, on account of its focal position between conscious and unconscious, ensures for the personality a new and more solid foundation. (p. 219)

No individual is born with the self. The self gradually develops out of inner conflicts, in the years of trial and experience. Jung rejected Freud's theory of developmental stages but suggested dividing the span of human life as follows:

The first five years of life are the years of self-protection. The libido is invested in the growth and development of basic skills like walking, talking, etc., necessary for survival. Around the age of five libido flows into the sexual values, reaches its peak around the teen-age years, and leads the individual in choosing a mate, forming a family, and establishing himself in his breadwinning occupations. The individual is extraverted, vigorous, outgoing. In the late thirties or early forties great changes take place. The individual turns gradually toward spiritual and philosophical values; he becomes introverted and interested in religious, moral, and spiritual values.

Personality Structure

Jung's theory of the psyche consists of two spheres, the conscious and the unconscious, in both of which the ego participates in aiding their functions. These diagrams (Jacobi 1945) show the relationship of the ego between the two spheres, and how they supplement, complement, and compensate each other. The ego is thereby not only a function of consciousness, but a center of reference for unconscious and conscious elements (diagram 1).

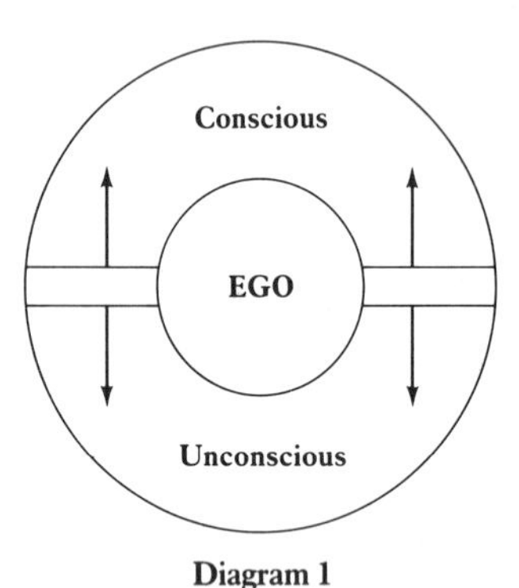

Diagram 1

The ego represents the total personality. Consciousness constitutes only a very small part of the psyche. It floats as a little island on the boundless sea of the unconscious (Jacobi 1945: 5). The ego, surrounded by and resting upon consciousness, is the part of the psyche concerned with adjustment to reality (diagram 2).

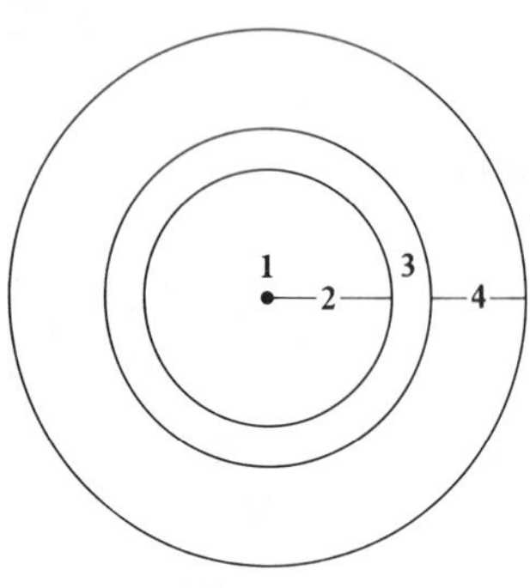

Diagram 2

1. Ego
2. Consciousness
3. Personal Unconscious
4. Collective Unconscious

Jung defined consciousness as "the function or activity which maintains the relation of the psychic contents to the Ego" (Jung 1933: 536). Diagram 3 shows the sphere of consciousness surrounded by unconscious factors.

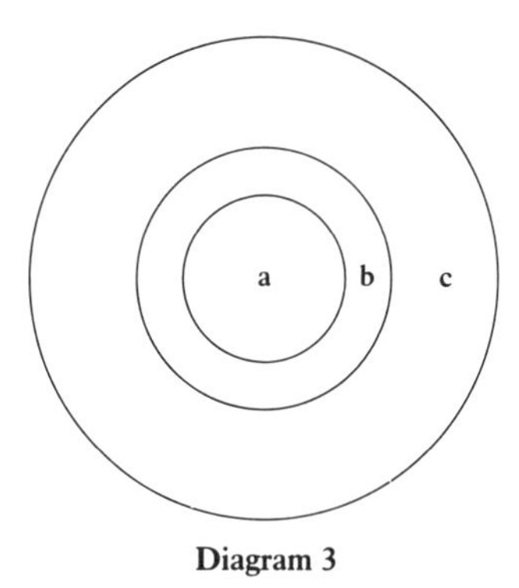

Diagram 3

a. Part of the collective unconscious that can never be raised to consciousness
b. Collective Unconscious
c. Personal Unconscious

Jung stressed individuality as one of the ego's main characteristics. The various elements which comprise individuality are the same, but they differ in clarity, scope, and emotional coloring. The ego is individual and unique, and it always retains a certain amount of its identity, although it undergoes some changes.

The integration of the contents of the collective unconscious must influence the ego. The assimilation of the archetypes adds not only to the field of consciousness, but also enhances the importance of the ego, especially when the ego itself has no critical approach to the unconscious. In some cases, the ego may become identical with the content of the archetypes which have been assimilated. For instance, a masculine consciousness may come under the influence of the anima and even become possessed by it. When the ego dominates the collective unconscious, one's self-confidence is increased and the individual believes he knows everything, and feel *au fait* with his unconscious. Some people become depressed, especially when their ego is dominated by the unconscious. In such a case, the individual loses his self-confidence, and submits to the unconscious. In the former case, the individual believes that his ego can fully control his unconscious. In the latter case, the individual tends to give up his sense of responsibility and feels that his ego is powerless against the unconscious forces.

Jung's theory operates with dialectic opposites, the conscious and unconscious playing the opposite yet complementary roles. Thus behind the overt self-confidence lies the feeling of helplessness, and the overt exuberance tries unsuccessfully to mask the hidden feelings. The reticence of a shy person serves as cover for the urge for power and this hidden forcefulness far exceeds the pretended self-confidence of the persona of the first type.

Ego and Persona

A well-functioning persona is a prerequisite of mental health; it is also a necessary element in forming successful relations with one's environment. However, a lasting maladjustment caused by an overidentification with one's persona, and especially with the persona's attitudes that strongly disagree with one's true ego, must lead to mental disturbances and severe neuroses.

Diagrams 4 and 5 portray the structure of the individual total psychic system. In diagram 4, the lowest circle is the largest. It rests beneath the

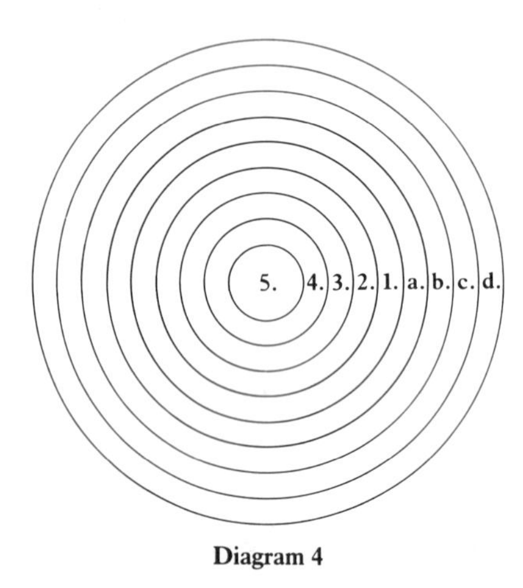

Diagram 4

1. Forgotten material
2. Repressed material
3. Emotions
4. Deep unconscious eruptions
5. Material unable to become conscious

a. Sensation
b. Feeling
c. Intuition
d. Thought

others, with the ego at the top. In diagram 5, at the bottom lies the central force out of which each individual psyche is differentiated. Each section of the diagram represents additional elements of the collective psyche.

The persona gradually becomes identified with the ego, but not with the entire self. The ego is a complex of representations which compose the center of the field of consciousness and possesses a high degree of continuity and identity. The ego occupies the center of the conscious. The identification of the persona with the ego is called the ego complex. "The persona represents the conscious attitude, and as such it is placed in the psyche as an opposite to the unconscious" (Progoff 1953: 85). The contents of the persona are likely to clash with one's unconscious and any

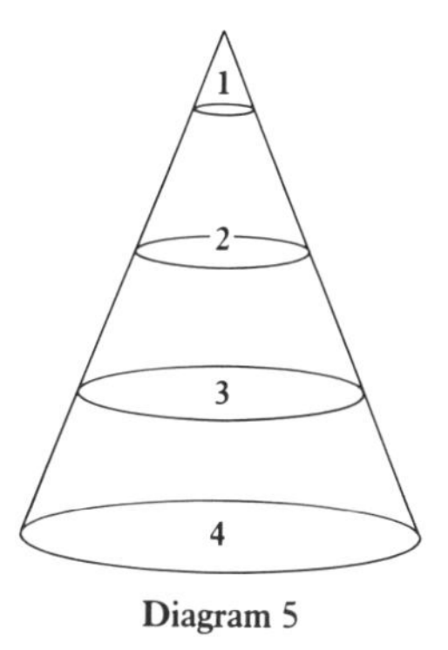

Diagram 5

1. Ego
2. Consciousness
3. Personal Unconscious
4. Collective Unconscious

extreme feature of the persona finds an extreme opposite in the unconscious. The persona must somehow be reconciled with the unconscious, or a vehement conflict of opposites may disturb the total personality.

The ego tends to foster the more developed and stronger components of personality and to absorb them in its conscious attitudes and its persona mask. The weaker parts of personality tend to be abandoned and fall back into the unconscious.

The Shadow

The negative complex of the personal unconscious, called the shadow, tends to appear on the surface in the conscious, and impose its will on the conscious parts of the personality. "It is as though a separate individual were in the personality deliberately doing things wrong against the wishes of the conscious ego" (Progoff 1953: 87).

In some cases the shadow shifts to the level of collective unconscious and joins the anima-animus archetype. In such a case, the shadow becomes expressed by a figure of the opposite sex. When the shadow reaches the collective unconscious, it becomes converted into a figure which is the opposite of the consciousness. The shadow of a man on the level of personal unconscious is expressed by a masculine figure, but the same shadow on the collective unconscious level becomes a feminine figure, and the shadow of a woman emerges from the collective unconscious in the shape of a man. The shadow can transform an animus to an anima and vice versa.

Becoming conscious involves individuation and drawing distinctions. The process of individuation implies becoming one's own self, and fulfillment of one's own specific nature, which includes some degree of egoism and individualism. The process of individuation is a conscious process. Its progress reduces inner tension by bringing together the opposites.

Fundamental Functions

The life energy, called libido, can be activated in "rational" processes determined by what Jung called "objective values," that is, activities which may be verifiable in terms of logical analysis. Or it may take the form of "irrational" processes, determined chiefly by "accidental perceptions," chance, and more or less illogical associations. Both cases are manifestations of the movements of the libido.

The rational process, in turn, is divided into two fundamental functions: "thinking" and "feeling." In parallel fashion, the irrational is divided into "sensation" and "intuition." The former functions are dominated by reasoning and judgment, the latter by intensity of perceptions but not by rational judgment. *Sensation* is the first reaction of the individual to the outer world; then comes *thinking*, or "interpretation of that which is perceived"; next comes *feeling*, or "evaluation of the perceived object"; last comes *intuition*, or the "immediate awareness of relationships." Furthermore, "sensation establishes what is actually given, thinking enables us to recognize its meaning, feeling tells of its value, and finally intuition points to the possibilities of the whence and whither that lie within the immediate facts. In this way we can orientate ourselves with respect to the immediate world as completely as when we locate a place geographically by latitude and longitude" (Jung 1933: 107).

Thinking and sensation are masculine personality traits, while intuition and feeling are feminine, but each individual is capable of all four functions. Uusually one of them is *dominant* in a given individual. The dominant function is carried by a great load of the libido energy into the conscious part of his personality and eventually, through fusion with his ego, becomes the guiding principle in his life. The entire mental life of an individual revolves around the dominant function. For example, all men use feeling for setting up the evolution criteria, but the "feeling type" will relate his entire life to this function. The dominant function occupies the center of the conscious, and its opposite, in this case "thinking," becomes necessarily unconscious. Any function on the unconscious level is undifferentiated, merged with other functions, and diffused. When the function is carried by the libido from the unconscious into the conscious it becomes *differentiated,* i.e., purified, separated from the opposite function. Usually the function which guarantees the most success, a talent, becomes dominant and differentiated.

The opposite function loses its load of energy to the dominant function. If, as a result of the drain of energy by the dominant function, the opposite function becomes exceedingly impoverished, it goes deep down to the lowest levels of the unconscious, stirs the archetypes, and leads toward mental disorder. The autonomous partial systems will then take possession of the human mind.

Personality Types

> When the orientation to the object and to objective facts is so predominant that the most frequent and essential decisions and actions are determined not by subjective values but by objective relations, one speaks of an extraverted attitude. When this is habitual, one speaks of an extraverted type. If a man so thinks, feels and acts, in a word, so *lives,* as to correspond *directly* with objective conditions and their claims, whether in a good sense or ill, he is extraverted.
>
> His entire consciousness looks outwards to the world, because the important and decisive determination always comes to him from without. But it comes to him from without only because that is where he expects it. (Jung 1923: 417)

In extraversion the libido moves toward the outer world. Accordingly, the extravert is guided by the impression that the external world leaves upon him. All interests, values, and attitudes are directed toward his physical and social environment.

The introvert, on the contrary

> selects the subjective determinants as the decisive ones. This type is guided, therefore, by that factor of perception and cognition which represents the receiving subjective disposition to the sense stimulus. . . . Whereas the extraverted type refers pre-eminently to that which reaches him from the object, the introvert principally relies upon that which the outer impression constellates in the subject. (Jung 1923; 472)

Extraversion and introversion can be organized around one of the fourfold features: thinking, feeling, sensation, and intuition. A brief description of each type of individual follows:

1. Extraverted Thinking Type. Accepts the world from the senses and uses his sensory impressions as a basis for logical analysis and construction of his reality. He is concerned with facts and their classification.
2. Extraverted Feeling Type. Determined by the feeling for the external object. The individual tends to feel and act according to the demands and expectations of the situation; he is able to establish friendship with others.

3. Extraverted Sensation Type. Conditioned by and oriented to the sensory and/or concrete features of a given object; he is the "realist" and "materialist."

4. Extraverted Intuition Type. The external object does not so much control his perception or sensation as offer him a suggestion for elaborating the possibilities of the object at the moment as something to manipulate and control.

5. Introverted Thinking Type. Marked by ideational patterns which have been almost completely organized subjectively till they suit the individual (so that he tends to become indifferent). He may have some success in social contacts.

6. Introverted Feeling Type. Dominated also by the "subjective factor." This individual lives within his own internal world of emotions and feelings. He is the daydreamer or the silent person who is at peace with the world.

7. Introverted Sensation Type. Though attending to the external world, his perceptions are dominated by his subjective internal state. The creative artist may have this kind of character makeup.

8. Introverted Intuition Type. Directs his attention to imagery. These images are clues to his activity. He lives within himself and may be the so-called dreamer, religious prophet, fanatical crank, or artist.

Jung stated that type differences can be modified, as when a natural-born introvert is forced by circumstances into extraversion, but he believed that such transpositions are rather superficial.

Between these two types lie most people, who display both extraversion and introversion. They are called *ambiverts*. According to Jung, the four functions of thinking, feeling, sensation, and intuition appear in all individuals, but one of these modes predominates. A particular person's response to the world and to himself might be said to be typically or characteristically in this or that modality, which does not imply an absence or the other features. Which of these extremes one should strive for is a question of norms and values. In our society, despite the culturally approved stimuli to become extraverted, we find a place for some introverts—though we are likely to consider them a bit "queer" or perhaps divergent from the "best" in our values; doubtless the majority of people fall into the middle range of so-called ambiverts.

Individual and Society

The plight of the individual in a mass society is that he is losing his individuality to the masses. The world is divided into two halves, each with certain subversives who have nothing to stop them in their spread of ideas except for the fairly intelligent, mentally stable portion of the population. An optimisitic view puts the limit of the number of these people at about 40 percent. The mass is crushing out the insight and reflection found in the individual, leading to doctrinaire and authoritarian tyranny, if ever the constitutional state should succumb.

For every manifest case of insanity there are at least ten latent cases who seldom get to the point of breaking out openly, but who are under the spell of morbid and perverse factors. Their ideas, brought out into the open by fanatical resentment, appeal to the irrationality of the masses and find fruitful soil, for they express resentments and motives which lurk in the minds of a great many people.

Jung's sociopsychological essay *The Undiscovered Self* (1959) describes the dangers of a "psychic epidemic." The banding together of individuals causes the extinction of the individual personality, making the group succumb easily to a dictator. Everything is dependent ultimately upon the individual but the habit of our age is of thinking only in terms of large numbers and mass organizations without the least consciousness that these organizations can be powerfully handled by one ruthless dictator. The individual becomes morally and spiritually inferior in the mass. The salvation of the world consists of the salvation of the individual soul. According to Jung, the resistance to the organized mass can be effected only by the man who is as well organized in his individuality as the mass itself.

A "psychic epidemic" can be initiated by a surprisingly small portion of society—these are the people who are not emotionally stable, but yet not so neurotic as to endanger their fellow men except in times of great stress. These are the people who "seldom get to the point of breaking down openly but whose views and behavior, for all their appearance of normality, are influenced by unconsciously morbid and perverse factors." And when the population as a whole finds itself in an atmosphere of general "collective possession," the latent neurotics and psychotics are in their element; for in such an agitated atmosphere their distorted views of reality can take root and flourish. These persons find that their ideas will be listened to, and, perhaps, accepted by the general populace, who are in a state of fear and who, in their desperation, will grasp eagerly at a promise of relief. The more normal people who usually would dismiss fanatical oratory as such are now able, under stress of fear not balanced by reason, to place their hopes in the nebulous guarantees of the circumstance-fortified fanatic. It is for lack of understanding and self-knowledge that these otherwise normal people are led astray.

In mass societies the State takes over the role of God. But the religious function cannot be done away with without secret doubts, which are immediately repressed in order to avoid conflict with the trend toward mass-mindedness. The result is a fanaticism which is used to stamp out the least flicker of opposition. Free opinion is suppressed and moral decision ruthlessly done away with, on the plea that the end justifies the means, even the vilest. Policy of the State is raised to a creed and the leader becomes a demigod and his followers heroes, martyrs, apostles, and missionaries. The State, like the church, demands enthusiasm, self-sacrifice, and love, and since religion presupposes "fear of God," the dictator-state manages to provide its own terror.

Religion

Religion means dependence upon and submission to irrational facts of experience. Religion allows the individual to exercise his own judgment and power of making decisions. It teaches of another authority opposed to the "world." The doctrine of the individual's dependence upon God makes such a claim upon him as to rival the world's claim. Religion estranges him from the world in the same way that he is estranged from himself when he succumbs to the mass mentality.

Religion recognizes the process of individuation as the incarnation and revelation of God, but contemporary scientific psychology proceeds abstractly and is often apt to disregard individual characteristics in its search for general knowledge. But even the churches themselves are apt to fall carelessly into the trap of mass manipulation. The churches attempt to gather individuals into a social organization of believers. Although a great social service is thus performed, it is as the expense of individual salvation. As a result, the individual must remain evermore watchful, for a favorable external environment strengthens the tendency to accept the community values as the only true ideals. In allowing the "authorities" to make his decisions, the individual forfeits his right, as well as his ability, to make his own. The more tasks the individual gives to higher social or political authorities, the more helpless and dependent he becomes.

The breaking down of religion into pseudoscience—the teaching of principles of faith as historical events subject to investigation—has wrought the individual with confusion. He is torn between blind belief and flat refusal. Natural human inclinations towards religion are being denied him, and he has no Faith. Since mysticism is a natural function of the human mind, the absence of true religion has given rise to new deities for worship: in the Western world these deities take the form of wealth, power, and so on.

Inner Conflicts

The conflict between allegiance to self and allegiance to external authorities has brought about a schism in human minds. Contemporary man, dimly aware that he is troubled within, calls upon the psychiatrist to cure him of his ills. The personal symptomatology of the patient is then referred to as "neurotic," because it usually deals with infantile fantasies that make an appearance in the adult consciousness. Actually, these "neurotic" tendencies are present in all persons, only manifesting themselves when abnormal or greatly disturbing conditions have upset the psychic balance of the individual as a child. Generally, neurosis makes itself seen when the (previously disbalanced) adult finds himself in a situation with which he cannot cope using his conscious powers, thereby admitting into his consciousness a flood of wish-fantasies in which the problem is solved, or at least delayed.

Infantile fantasies, however, are not abnormal at all. The child creates fantasies and other make-believe phenomena as exercises against the time when his imaginative powers will become necessary for his livelihood or his survival. Such mental mechanisms have great survival value and are deeply rooted in instinct. Human survival consists primarily in adaptation to existing conditions. No less important than man's need to adapt to a changing world is his need to adapt mentally as an individual so that he may withstand the weathering forces of society.

The present-day philosophy of life fails to offer a way of life. In modern times, philosophy has become an academic issue, not related to real life. The religious beliefs practiced in the Middle Ages have come under harsh scrutiny and criticism in the light of scientific progress. An enormous gap has emerged between faith and knowledge. Even the churches are attempting to "demythologize" the events pertaining to their faith, and thus render them objects of knowledge.

The rupture between faith and knowledge is a symptom of the split consciousness which is so characteristic of the mental disorder of our day. Viewing society as a macrocosm of the individual, one could say that the former is suffering a neurosis in much the same manner as an individual patient. One of the manifestations of this neurosis is the substitution of the symbol for the reality. In one such case, we observe a substitution and deification of the word as the reality. Words like "society" and "state" have been concretized almost to the point of personification. Removed further and further from its referent, a word like "state" becomes a reality; and credulity in the abstraction as the reality becomes a tool in the hand of mass manipulators.

A typical conflict can be described as the clash between socialization (species preservation) and individuation (self-preservation). Man's innate instincts are being overridden by his supreme ability to reason and modify his environment. In creating a society, he is forced either to sublimate or to repress his instincts.

Modern man is in the throes of a serious inner conflict which could be solved through self-knowledge. Self-knowledge, through the exploring of one's own soul, brings to light our instincts and their world of imagery throws some light on the powers of the unconscious. Whoever has insight into his own actions may also find an access to his unconscious, and thus involuntarily exercise some control over his fate and environment.

Alchemy

Jung viewed alchemy as a symbolic presentation of the process of individuation. The turning of base metals into noble ones reflected the human striving towards an ideal. Mandala, the main symbol of alchemy, represents man's quest for wholeness and perfection. Mandala was the archetype of inner order (Evans 1964: 62–63). Mandala is an ancient symbol that appears in dreams, art, religion, and mythology. Jung believed that mandala symbols were not created by a particular religious sect but originated in the unconscious, in dreams and visions. Mandala is one of the oldest and most universal alchemic and mythological symbols.

Burt (1964: 478–479) hypothesized that Jung believed himself to be a possessor of supranatural powers, capable of mystical experiences. While this hypothesis is a matter for biographers, Jung's interest in mythology merits attention as it represents a branching of the psychoanalytic idea of unconscious processes.

Mythology and Religion

According to Jung, primitive man was not much interested in objective explanation of the obvious, but his unconscious psyche had an irresistible urge to relate his sensory perceptions to his inner, dimly perceived experiences. It was not enough for the primitive to see the sun rise and set; this external observation was transformed into a psychic happening; the sun was made to represent the fate of a god or a hero, who dwells in the soul of man. The regular processes of nature, such as summer and winter, the phases of the moon, and the rainy seasons, were symbolic expressions of the inner unconscious drama of the psyche which becomes accessible to the consciousness through the mechanism or projection. Primitive men channeled archetypes into sacred and dogmatic images, and developed complex rituals aimed at the assuaging of the personified images they themselves had created. Primitive men defended themselves against their own projection by a series of rites such as exorcising of spirits, lifting of spells, and averting of the evil omen. Christianity channeled the archetypes into a series of symbols, rituals, and sacred images. People ascribe meaning to these images until they discover, as it had happened during the Reformation, that they do not have any meaning but that the human mind has created them. However, as soon as the people discard the dogma for archetypal images, they lose their emotional balance and feel confused. Sometimes, when a whole nation throws away the religious and moral norms and rites as being stupid, the animalistic drives break loose, and a wave of demoralization and depravation floods the civilized nation. Thus the fear of the unconscious breaking loose is noticeable in the constraints of the Catholic religion.

Primitive men erected powerful archetypical images to ward off the threats of their own unconscious, which they feared. But as soon as the unconscious touched primitive man, he became unconscious of himself. However, at the early stage the conscious was rather childish and unstable. A wave of the unconscious could have easily overcome the thin layer of conscious. Primitive men were afraid of uncontrolled emotions and erected walls and dams of rites and dogmas.

According to Jung, religion has two aims. First, religion represents the human endeavor aimed at the creation of a "god-figure" as an ideal self. The god-figure is a symbol of the ideal self and it represents more than the self. The Christ figure is a good case in point, for it does not contain the "nocturnal side of the psyche's nature," and is without sin. In the formation of religious symbols, Jung also refers to the symbol as a "libido analogue" whose function it is to transform instinctual energy into useful cultural activities (Jung 1958: 156).

> How are we to explain the religious processes, for instance, the nature of which is essentially symbolical? In the form of representations, symbols are religious ideas; in the form of action they are rites or ceremonies. Symbols are the manifestation and expression of the excess libido. At the same time they are transitions to new activities, which must specifically be characterized as cultural activities. (Jung 1928: 53)

Thus religion represents cultural values and also the ideal self.

Western man has practically lost the great heritage of Christian symbolism. A wholesome development of personality requires a harmonious balance between the conscious and unconscious elements, but unfortunately, Western man suppressed the individual symbol making and impoverished his psyche.

Between East and West

Jung tried to build bridges between Western and Eastern thought. Jung was especially impressed by the Chinese philosophy and found the Chinese texts to be not a sort of "mystical intuitions of pathological cranks," but realistic insights which "grew logically and organically from the deepest instincts, and which for [Westerners] is forever inaccessible and impossible to imitate" (Jung 1938: 7–8). Small wonder that the Europeans and Americans have failed to grasp the full meaning of the Chinese word *Tao*. Some of them call it "the way," "providence," "God," or "meaning." The original Chinese symbols for Tao are "head" and "going." According to Jung, the head represents consciousness, and going means travel. Thus Tao means "the conscious way." Human nature and life, according to the test, are part of "the light of Heaven," and light is a symbol of consciousness. In the *Hui Ming Ching*, the terms human nature and consciousness are used interchangeably. Jung maintained that "the Tao is the method or conscious way by which to unite what is separated," namely life and consciousness.

According to Jung, the union of opposites has been expressed in mythological symbols as well as in the development of individuals. When individuals give vent to their fantasy, intuitive formulations of unconsciously experienced principles take on distinct shapes of persons or actions. When one tries to draw these fantasies, they usually take on the shape of a magic circle, called mandala.

Mandala drawings are frequently made by mental patients who are totally unaware of any leaning or connection with a religion. Some of Jung's female patients did not draw mandalas but danced their shape. The Chinese mandala, the Golden Flower, symbolizes light, and Tao is the heavenly light. According to Jung, "every separate thought takes shape and becomes visible in color and form. The total spiritual power unfolds its traces" (1954–1971, 13: 76ff). However, should one remain in a position of contemplation as described and shown by diagrams in the *Hui Ming Ching*, one would experience schizophrenia. Jung believed that the Chinese sages anticipated this danger and they warned: "The shape formed by the spirit-fire are only empty colors and forms. The light of human nature shines back on the primordial, the true." Thus, Jung maintained, the magical circle was drawn to protect the unity of consciousness from the unconscious, and by describing the figures of thought as "empty colors and forms," the Chinese sages protected the conscious and the unconscious (Jung 1938: 29).

According to Jung, the Chinese philosophy is not logical in the Western sense of the word but rather a set of intuitive ideas. One can comprehend their true meaning only through their symbolic representations. Jung defined the anima as a personification of the unconscious in general, or as a bridge to the unconscious. The Western way of thinking is based on the vantage point of the conscious, but the Eastern thinking takes its roots from the unconscious and assumes that all that is conscious originates in the unconscious. Jung wrote in *Psychology and Religion: West and East:*

> In the East there is an abundance of conceptions and teachings that give full expression to the creative fantasy; in fact, protection is needed against an excess of it. We, on the other hand, regard fantasy as worthless subjective day-dreaming. Naturally, the figures of the unconscious do not appear in the form of abstractions stripped of all imaginative trappings; on the contrary, they are embedded in a web of fantasies of extraordinary variety and bewildering profusion. The East can reject these fantasies because it has long since extracted their essence and condensed it in profound teachings. But we have never experienced these fantasies, much less extracted their quintessence. We still have a large stretch of experience to catch up with, and only when we have found the sense in apparent nonsense can we separate the valuable from the worthless. We can be sure that the essence we extract from our experience will be quite different from what the East offers us today. The East came to its knowledge of inner things in childlike ignorance of the external world. We, on the other hand, shall explore the psyche and its depths supported by an immense knowledge of history and science. . . . We are already building a psychology . . . that gives us the key to the very things that the East discovered. (1938: 43)

Jung vs. Intellect

Jung accused Western civilization of being one-sided and intellectual and of neglecting the emotional and intuitive aspects of human personality. He wrote in *The Commentary on the Secret of the Golden Flower* (in *Alchemical Studies*) as follows:

> For a long time the spirit, and the sufferings of the spirit, were positive values and the things most worth striving for in our peculiar Christian culture. Only in the course of the nineteenth century, when spirit began

> to degenerate into intellect, did a reaction set in against the unbearable dominance of intellectualism, and this led to the unpardonable mistake of confusing intellect with spirit and blaming the latter for the misdeeds of the former. The intellect does indeed do harm to the soul when it dares to possess itself of the heritage of the spirit. It is in no way fitted to do this, for spirit is something higher than intellect since it embraces the latter and includes the feelings as well. It is a guiding principle of life that strives towards superhuman, shining heights. Opposed to this *yang* principle is the dark, feminine, earthbound *yin*, whose emotionality and instinctuality reach back into the depths of time and down into the labyrinth of the physiological continuum. No doubt these are purely intuitive ideas, but one can hardly dispense with them if one is trying to understand the nature of the human psyche. The Chinese could not do without them because, as the history of Chinese philosophy shows, they never strayed so far from the central psychic facts as to lose themselves in a one-sided over-development and over-valuation of a single psychic function. They never failed to acknowledge the paradoxicality and polarity of all life. The opposites always balanced one another—sign of high culture. One-sidedness, though it lends momentum, is a mark of barbarism. The reaction that is now beginning in the West against the intellect in favor of feeling, or in favor of intuition, seems to me a sign of cultural advance, a widening of consciousness beyond the narrow confines of a tyrannical intellect. (Jung 1954–1971, 13: 9)

And this was, perhaps, the keynote in Jung's message.

BIBLIOGRAPHY

Adler, G. 1967. *Studies in Analytical Psychology*. New York: Putnam.

Burt, C. 1964. Baudouin on Jung. *British Journal of Psychology*, 55: 477–484.

Evans, R. 1964. *Conversations with Carl Jung*. Princeton, N.J.: Van Nostrand.

Friedman, P. and J. Goldstein. 1964. Some comments on the psychology of C. G. Jung. *Psychoanalytic Quarterly*, 33: 194–225.

Jacobi, J. 1945. *The Psychology of Jung*. New Haven: Yale University Press.

—— 1959. *Complex, Archetype, Symbol in the Psychology of C. G. Jung*. New York: Pantheon.

Jung, C. G. 1923. *Psychological Types*. London: Routledge and Kegan Paul.

—— 1928. *Contributions to Analytic Psychology*. London: Routledge and Kegan Paul.

—— 1954–1971. *Collected Works*. Princeton, N.J.: Princeton University Press.

Vol. 1: *Psychiatric Studies*. 2d ed. 1970.

Vol. 2: *Experimental Researches*. 1971

Vol. 3: *The Psychogenesis of Mental Disease*. 1960.

Vol. 4: *Freud and Psychoanalysis*. 1961.

Vol. 5: *Symbols of Transformation*. 2d ed. 1967.

Vol. 6: *Psychological Types*. 1971.

Vol. 7: *Two Essays on Analytical Psychology*. 2d ed. 1966.

Vol. 8: *The Structure and Dynamics of the Psyche*. 2d ed. 1968.

Vol. 9: Part I. *The Archetypes and the Collective Unconscious*. 2d ed. 1968.

Vol. 9: Part II. *Aion: Researches into the Phenomenology of the Self*. 2d ed. 1968.

Vol. 10: *Civilization in Transition*. 2d ed. 1970.

Vol. 11: *Psychology and Religion: West and East*. 2d ed. 1969.

Vol. 12: *Psychology and Alchemy*. 2d ed. 1968.

Vol. 13: *Alchemical Studies*. 1968.

Vol. 14: *Mysterium Coniunctionis*. 2d ed. 1970.

Vol. 15: *The Spirit in Man, Art, and Literature*. 1966.
Vol. 16: *The Practice of Psychotherapy*. 2d ed. 1966.
Vol. 17: *The Development of Personality*. 1954.

—— 1955. *Modern Man in Search of a Soul*. New York: Harcourt, Brace, Jovanovich.

—— 1961. *Memories, Dreams, Reflections*. New York: Random House.

—— 1964. *Man and His Symbols*. New York: Doubleday.

—— 1968. *Analytical Psychology: Its Theory and Practice*. New York: Random House.

Jung, C. G. and C. Kerenyi. 1949. *Essays on a Science of Mythology*. New York: Pantheon.

Progoff, I. 1953. *Jung's Psychology and Its Social Meaning*. New York: Julian.

[SIXTEEN]

Horney's "New Ways" in Psychoanalysis

KAREN Horney's theory represents a distinct rebellion against Freud's system. Horney accused Freud of being biologically oriented and relating psychological processes to chemical and physical forces as if the human mind were a derivative of biochemical factors. Horney maintained that Freud's theory was reductionistic. Although, she admits, Freud "encouraged the venture into a psychological understanding of phenomena which hitherto had been ascribed to organic stimuli" (1939: 18). Ultimately, "Freud's concepts of libido and id are rooted in organic factors." Horney's theory is radically nonreductionistic, and her models of normal and disturbed personality are free from any organic elements; they are wholly psychogenic.

Horney did not share Freud's views concerning sexuality. The basic contention implicit in Freud's libido theory, Horney wrote, "is that all bodily sensations of a pleasurable nature, or strivings for them, are sexual in nature" (1939: 50).

She substituted the sexual drive by the need to be accepted.

> Horney's growth-oriented, holistic, process-patterned system thinking is a rejection of the repetition compulsion, fixation, regression, the id and death instincts. "Freud's mechanistic evolutionistic thinking . . . implies that present manifestations not only are conditioned by the past, but contain nothing but the past; nothing really new is created in the process of development; what we see today is only the old in a changed form (1939: 42). What Freud regarded as secondary process thinking, namely, creativity through sublimation, Horney regarded as primary, and what Freud called primary process thinking, Horney saw as a secondary consequence of growth blockage and distortion. She discarded the notion of the Oedipus complex but retained "the highly construc-

tive finding that early relationships in their totality mold the character to an extent which can be scarcely overestimated" (1939: 87) but warned against a "one-sided fascination" with childhood.

We see unconscious processes as not directly observable, but inferred, and recognize that while the patient does not know, on another level he "knows that he knows" (Freud). We do not make unconscious processes identical with the prerationative levels of our postulated symbolic spiral because of the way we defined them and because of the holistic theory into which they fit. We do feel that these levels are analogous with and a close approximation to unconscious processes. We also feel that the notion of the symbolic spiral, which participates in defining the actual situation, communing and communicating, and the doctor-patient relationship, gives a much more comprehensive basis for mutual contacting of therapist and patient through having wider and deeper access to each other. (Kelman and Vollmerhausen 1967: 387–388)

Motivation

The need to be accepted is the fundamental driving force in human beings. The search for pleasure is not the only guiding force. In some instances, the need to be accepted and to feel safe may outdistance the search for pleasure. "Man is ruled not by the pleasure principle alone but by two guiding principles: safety and satisfaction," says Horney. This statement is the motto of all her writings and the cornerstone of her theories. "People can renounce food, money, attention, and affection so long as they are only renouncing satisfaction, but they cannot renounce these things if without them they would be or feel in danger of destitution or starvation or of being helplessly exposed to hostility, in other words, if they would lose their feeling of safety."

This is the core of Horney's theory. Each individual has certain fundamental needs, for food, rest, and sex, and these needs must be satisfied. They cannot be combined into one, e.g., sex, as Freud combined them. However, all these needs can be brought together under the common heading of seeking satisfaction, and they represent the principle of pleasure.

Despite the fact that food and sex are the primary needs, they are not the decisive factors in human behavior. People can renounce these things, states Horney emphatically. They will do so if exposed to danger.

If this is the case, what is the decisive driving power? Horney would say: *the need for safety*, the need to be secure and free from fear.

Fear and safety, these are the two poles of the basic needs. Man needs safety and avoids fear. He cannot enjoy satisfaction of needs unless he feels safe. Fear is the greatest enemy of man's health and happiness, and search for safety is the guiding principle in human behavior.

Fear and Anxiety

Horney distinguishes between fear and anxiety. Fear is an emotional reaction to a real danger, while anxiety is a reaction to a situation perceived subjectively as dangerous. Lack of acceptance in childhood creates the basic anxiety. "According to this concept the child not only fears punishment or desertion because of forbidden drives, but he feels the environment as a menace to his entire development and to his not legitimate wishes and strivings."

Lack of satisfaction produces anxiety, said Freud. Horney says: Lack of acceptance produces *basic anxiety*. Basic anxiety is not innate; it is a result of environmental factors. The rejected or unwanted child, or a child in a broken or hostile home feels that he is "being isolated and can produce this insecurity in a child." The driving force in man is postulated by Horney is a noninstinctual manner; fear and anxiety are basic emotions though the result of life experiences.

Here lies one of the fundamental differences between Freud's system and Horney's. Freud regarded love and hatred as basic emotions while anxiety is a secondary phenomenon produced by a thwarting of the basic desires. Fear is not innate; it develops as a result of coping with reality and must be related to the reality-oriented ego. The counterpart to love is hatred, and fear develops later on.

Horney regards anxiety as a basic feeling and a counterpart to love. People need to be accepted, and basic anxiety is a person's reaction to lack of acceptance.

A child who is accepted and safe in his home environment develops the feeling of *basic security*. Basic security, produced by a favorable environment, fosters natural and successful growing.

> Implicit in Horney's holistic concepts is the assumption that there is no life course in which every developmental experience has been traumatic nor one from which all deleterious influences have been absent. As a result of harmonious constructive experiences on one hand and the destructive traumatic experiences on the other, the personality of the child develops around two nuclei and forms two basic patterns. In the one there is a basic feeling of confidence that one's striving for love and

> belonging as well as autonomy may be more or less realized. In the other, there is a basic feeling of helplessness and isolation in a potentially hostile world, which feeling would undermine the striving for autonomy and belonging.
>
> The pattern that is most deeply and extensively embedded determines how the energies, abilities, and resources of the individual will be used. These may serve predominantly to support and maintain the defensive systems, or largely the constructive resources. These patterns proceed in changing proportions and in an oscillating manner. Adverse environmental and internal factors propel the development of the defensive systems, restrict growing ones, extend and embed basic anxiety, which is inextricable with basic hostility (1937: ch. 3 and 4). (Kelman and Vollmerhausen 1967: 381)

Horney espouses an optimistic view on human mature. Human beings are naturally disposed to a wholesome growth and fulfillment of their inherent potentialities. Unless exposed to adverse conditions in childhood, people tend to grow towards harmonious life, self-fulfillment, and happiness. Horney wrote: "My own belief is that man has the capacity as well as the desire to develop his potentialities and become a decent human being. I believe that man can change and go on changing as long as he lives" (1945: 19).

Horney rejected Freud's theory of innate instincts, both the early version and the final two-factor Eros and Thanatos theory. Horney was critical of what she believed to be Freud's undue pessimism concerning human nature. She totally rejected the idea of the death instinct and innate self-destructive forces.

According to Horney, personality is shaped by experience, especially by interaction with one's family environment. A home which provides for safety and gives the child the feeling of being accepted, protected, and free from fear fosters normal and wholesome development.

Fear and safety are the two fundamental issues in human growth. People avoid fearful situations and seek safety. Fear is the archenemy of mental health, and safety is its best protector.

One must distinguish fear from anxiety. Horney defines fear as a reaction to a realistic threat and anxiety as a reaction to a situation subjectively perceived as threatening. A child who does not feel accepted by his parents may overreact to their threats of punishment or desertion and feel menaced in his very existence and fears retaliation not only for his violation of parental prohibition but even for his hidden wishes.

Normal Development

Horney rejected Freud's theory of psychosexual developmental stages. Horney's view comes close to Adler and Sullivan, for she emphasizes the role of environment. According to her, the particular needs which are relevant to understanding the personality and its difficulties are not instinctual in character but are created by the entirety of conditions under which we live (1939: 38).

The development of the child does not follow biologically determined Freudian phases but entirely depends on how the child is treated. The child may or may not develop oral or anal fixations, depending on the interaction with his parents. The Oedipus complex and any other mental complexes are artifacts of a particular culture and may develop as a result of a particular intrafamilial pattern of interaction.

Neurosis

In a normal environment (to Horney "normal" and "natural" are synonymous) the child can relate to parents, siblings, and everyone else in a "natural," spontaneous, free-floating way in accordance with his wishes and feelings. He relates to people in an open, honest, and basically friendly manner and stands up bravely for his rights. Self-realization, that is, the striving for a full use of one's potentialities is the main road of normal development.

According to Horney, neurosis is a self-perpetuating, unnatural process of prevented or distorted growing. The child's environment and the society at large breed neurosis by creating situations fraught with rejection, unfair competition, and insecurity. Our society is conducive to neurosis because it carries "the germs of destructive rivalry, disarrangement, suspicion, begrudging, envy into every human relationship" (Hroney 1937: 113).

Neurosis is not a result of a conflict between the id, ego, and superego, but it is produced by a conflict between the individual and his environment. In her early writings, Horney described neurosis as a certain "rigidity in reaction and discrepancy between potentialities and achievement" (1937: 28).

In a later work she described neurosis in terms of *basic anxiety*, as that which starts with

> . . . the feeling a child has of being isolated and helpless in a potentially hostile world. A wide range of adverse factors in the environment

> can produce this insecurity in a child: direct or indirect dominations, indifference, erratic behavior, lack of respect for the child's individual needs, lack of real guidance, disparaging attitudes, too much admonition or the absence of it, lack of reliable warmth, having to take sides in parental disagreements, too much or too little responsibility, overprotection, isolation from other children, injustices, discrimination, unkept promises, hostile atmosphere, and so on and so on." (1945: 41)

Anything that disturbs the security of the child in relation to his parents, anything that makes the child feel unwanted or rejected produces basic anxiety.

The insecure, anxious child develops strategies by which to cope with his feelings of isolation and helplessness; he may become hostile and seek to avenge himself against those who have rejected or mistreated him; he may become overly submissive in order to win back the love that he feels he has lost; he may develop an unrealistic idealized picture of himself in order to compensate for his feelings of inferiority (Horney 1950); he may also try to bribe others into loving him or he may use threats to force people to like him; he may wallow in self-pity in order to gain people's sympathy; he may seek to obtain power over others thus compensating for his sense of helplessness and providing an outlet for his hostility; he may become highly competitive; and, finally, he may turn his aggression inward and belittle himself.

Neurotic Trends

Any one of these particular strategies may become a more or less permanent fixture in the individual's personality. Horney presents a list of ten of these strategies or "needs" which are acquired as a consequence of trying to find solutions for the problem of disturbed human relationships (Horney 1942). The needs are called "neurotic" because they are irrational solutions to the problem. The ten neurotic needs are: 1) the neurotic need for affection and approval; 2) the neurotic need for a "partner" who will take over one's life; 3) the neurotic need to restrict one's life within narrow borders; 4) the neurotic need for power; 5) the neurotic need to exploit others; 6) the neurotic need for prestige; 7) the neurotic need for personal admiration; 8) the neurotic ambition for personal achievement; 9) the neurotic need for self-sufficiency and independence; and 10) the neurotic need for perfection and unassailability.

The three neurotic trends, namely, *compliance, defiance,* and *withdrawal,* may appear simultaneously, and the child feels caught between conflicting compulsory tendencies. This *basic conflict* may lead to repressions which do not solve the problem, but only deepen the feelings of inadequacy and insecurity.

Idealized Image

In many instances the individual unconsciously tries to overcome these feelings by creating a false picture of himself, called by Horney the *idealized image*. The individual begins to believe that his compulsive compliance is a sign of him being a noble character; his compulsive defiance is viewed as a sign of courage; and his compulsive withdrawal is perceived as an evidence of intellectual independence.

The neurotic individual tends to substitute *false pride* and attribute to oneself virtues he does not possess. The striving for *self-actualization*, that is, becoming one's idealized image, may take place instead of the healthy fulfillment of one's true values, called self-realization.

A well-adjusted individual can cope with life's inevitable moments of frustration and defeat, but the neurotic develops self-hate whenever he fails in attaining the glorious picture of himself. He tends to belittle his own achievement, develops a guilt feeling for noncommitted transgressions.

Self-hate and self-destructive behavior are the inevitable consequences of self-idealization and neurotic pride. Continuous self-coercion, called by Horney the *tyranny of the should* (Horney 1950), makes one a slave to his self-imposed, compulsive striving to perfection, in order to live up to his idealized image. Again, instead of the natural drive to fulfill his innate and acquired potentialities (called by Horney *self-realization*), the individual compulsively tries to be what he is not (self-actualization). Being unable to attain the false goal, the individual blames himself and is tormented by totally irrational feelings of guilt, hypochondrised fears, and, ultimately, driven by self-contempt and self-hatred, he may try to put an end to his life.

Major Solution

There are three major "solutions" to this inner conflict, the *expansive* solution, the *self-effacing* solution, and *resignation*. The expansive solution implies an unconditional acceptance of the idealized self. The striving for superiority motivates the individual's behavior. One may develop a *narcissistic* pattern of behavior, making oneself believe in one's superiority and expecting to be surrounded by a general and unswerving admiration. One may become a *perfectionist*, acting in a compulsive manner and rejecting any rational compromise. One may also become *arrogant-*

vindictive, repressing one's own need for dependence and showing no compassion for others.

The *self-effacing* attitude implies identification with failure to become the idealized self. The self-effacing individual believes himself to be an innocent victim and craves for love and compassion. He seeks relationships in which he could depend on others. Feeling unworthy or helpless, he embraces his own suffering and wallows in his love-demanding misery.

Resignation, the third solution, implies withdrawal.

In addition to these three major "solutions," an individual may develop several other defenses.

Foremost among these auxiliary defenses is the process of alienation from self. It is both an outcome of the neurotic development as well as a defensive measure to relieve tension (Horney 1950: ch. 6). In its more extreme form it appears as depersonalization. All that is compulsive moves the individual away from his real self as well as from his actual or empirical self. His possessions and experiences are not felt as his. He becomes numb and remote, a stranger to himself. His relations to himself and others become increasingly impersonal and mechanical. He loses the capacity to feel his own feelings, made worse by his neurotic pride dictating what he should and should not feel. He has lost his inner self-directedness and has become other directed.

Another defense against inner tension and conflict is the *externalization* of inner experience (Horney 1950: ch. 7).

> This is a more comprehensive phenomenon than projection, which is concerned with "the shifting of blame and responsibility to someone else for subjectively rejected trends or qualities" (Horney 1945: 116). Any psychic process, including aspects of the pride system and the real self, can be externalized. A further measure for tension reduction and an outcome of the preceding measures is *psychic fragmentation* or *compartmentalization*. When and as these fail to do away with disruptive conflicts, *automatic self-control* sets in to check all impulses indiscriminately. This results in an increased rigidity and constriction of the personality. The last auxiliary measure is the belief in the supremacy of the mind. It functions as a spectator of the self, gleefully and sadistically finding fault with it. It also functions as a magic ruler with beliefs in omnipotence and omniscience. (Kelman and Vollmerhausen 1967: 384)

Kelman's Contribution

Harold Kelman further developed Horney's psychoanalytic system. He introduced the concept of symbolic self as the constantly forming and changing process arising from the inner notions of identity, conceptions of body and sex, and of relations to other people and to the universe (Kelman 1971).

> What we start and end with, and are always in and of, is the actual, total, immediate, present situation. This means that "the only place we can be is here, the only time we can be is now, and the only" organismal-environmental happenings becoming manifest as "feelings, sensations, perceptions and thoughts, signs, gestures and sounds which we can be, no have, are here-now." These happenings may be clothed in symbols having the time form of past, present, or future, and the place form of here and there. These "forms will be emerging and being resorbed back into their source. . . . We have two modes of knowing these phenomena, inferentially through rationality, and directly through numinous awareness. . . . The sole aim of such knowing . . . is the widening and deepening of here-now experiencing of those symbol forms and what they point at, preformed. . . . This aim becomes more possible in and through the relating process during which more moments of communing are obtaining" (Kelman 1971).

And, furthermore, about the symbolizing process:

> The moment there is awareness of being and any of its attributes they have emerged into form. They have taken form through the forming process. Form may become organized as a sequential hierarchy of forms, denotable, conatable, and connotable. The forms may be vague, inchoate, increasingly defined, and variously named. . . .
>
> The symbolizing process, an aspect of human integrating, may be described metaphorically as a spiral or as a sequence of interconnected levels starting from the ground of all forms, phenomena, and appearances, variously named through time as *chit* (Hinduism), *hsing* (in Chinese philosophy), *tathata* (Zen), *pure fact* (Northrop 1948) and *dasein* (existentialism). At the bottom of the spiral is *pure fact,* from which all forms emerge and back into which they are resorbed. (Kelman and Vollmerhausen 1967: 385–386)

Kelman also introduced new ideas on dream interpretation. He defined the holistic process nature of technique, interpreting, and dreaming. He indicated the ways theory is integrated with techniques. Kelman attempted some limited elucidation of the phenomenologic and existential approaches. A variety of dreams, dream series, and issues relating to dreams were used to elucidate the process nature of these concepts.

BIBLIOGRAPHY

Horney, K., ed. 1946. *Are You Considering Psychoanalysis?* New York: Norton.

—— 1950. *Neurosis and Human Growth*. New York: Norton.

—— 1937. *The Neurotic Personality of Our Time*. New York: Norton.

—— 1939. *New Ways in Psychoanalysis*. New York: Norton.

—— 1942. *Self-Analysis*. New York: Norton.

—— 1945. *Our Inner Conflicts*. New York: Norton.

—— 1951. On feeling abused. *American Journal of Psychoanalysis*, 11: 5–12.

—— 1952. The paucity of inner experiences. *American Journal of Psychoanalysis*, 12: 3–9.

Kelman, H. 1971. *Helping People*. New York: Science House.

—— 1957. A unitary theory of anxiety. *American Journal of Psychoanalysis*, 17: 127–60.

Kelman, H. and J. W. Vollmerhausen. 1967. On Horney's psychoanalytic techniques: developments and perspectives. In B. B. Wolman, ed., *Psychoanalytic Techniques*, pp. 379–423. New York: Basic Books.

Martin, A. R. 1969. Idle hands and giddy minds. *American Journal of Psychoanalysis*, 29: 147–156.

—— 1972. Cultural impairment of our inner resources: an empirical inquiry. *American Journal of Psychoanalysis*, 32: 127–144.

Rubins, J. L. 1969, 1970, 1971. A holistic (Horney) approach to the psychoses: The schizophrenias. *American Journal of Psychoanalysis*, 29: 133–46; 30: 30–50; 31: 136–151.

Shainberg, D. 1973. *The Transforming Self*. New York: Intercontinental Medical Book Corporation.

Sheiner, S. 1969. The communications of psychotics. *American Journal of Psychoanalysis*, 29: 11–16.

—— 1968. The intensity of casual relationships in schizophrenia. *American Journal of Psychoanalysis*, 28: 156–161.

[SEVENTEEN]
Sullivan's Interpersonal Theory

HARRY Stack Sullivan developed his theory in the course of psychiatric work with severely disturbed patients. He was greatly influenced by George Mead's theory of social roles, Kurt Lewin's field theory, Adolf Meyer's biological approach, and Freud's theory of unconscious. Like Horney, Sullivan stressed the importance of acceptance and security in the upbringing of children, but in contradistinction to Horney, Sullivan's system is deeply rooted in psychophysiology.

Satisfaction and Security

Sullivan combined both the physiological and sociocultural aspects in his dualistic theory of motivation. According to Sullivan, *satisfaction* depends on organic elements, such as inner organs and their tissues. *Security*, however, depends on interpersonal relations.

Both the striped and the unstriped muscles serve one purpose: satisfaction. What we experience as thirst or hunger or as a desire for air or sex is a result of muscular contractions. Physiological tension "provokes" our pursuit of satisfaction, and satisfaction brings relaxation and relief of tension.

Tonic changes in the unstriped, involuntary muscles of the viscera—the internal organs of the body—are, from birth onward, intimately related to the experiencing of desires, needs for satisfaction. Heightened tone of the stomach wall is called out by depletion of our chemical supplies. Throughout life the pursuit of satisfaction is physiologically provoked by increased tone in some unstriped muscles; and the securing of the satisfaction is a relaxation of this tone.

The entire process of seeking satisfaction is interpreted in a reductionist manner. The organic state, such as a heightened tone of the stomach wall, causes us to experience a "desire." When satisfaction is secured, the muscles relax and our alertness is reduced. The organism tends to rest or sleep. The highest experienced tension is terror, the most profound satisfaction and relaxation is sleep.

This interpretation of human needs has little in common with Freud. It seems to be closer to some kind of organismic theory. Sullivan's theory invites comparison with Cannon's homeostatic system: *tension* and *relief* follow each other, especially since tension and relief are interpreted by Sullivan in purely physiological terms, as a contraction and relaxation of muscles. This physiological process is later modified by social influences. The child soon learns that on certain occasions the immediate relaxation of muscles meets with parental disapproval. Parental disapproval causes a feeling of discomfort. This *empathized discomfort* stems not from the organism but from *interpersonal relationships*. The discomfort is so strong that it destroys the original comfort of relaxation. For example, as soon as there is any tension in bladder or bowels, the proper muscles act immediately to bring relief. However, this automatic relaxation may invite parental hostility, which produces in the child a feeling of "empathized discomfort." This feeling will bring the child to learn "to suffer increasing tension in the bladder and rectum and to resist the automatic relaxation of the sphincter muscles concerned in retaining the urine and feces. Failures in this are often accompanied by empathized discomfort, and success is often the occasion of empathized comfort which is added to the satisfaction from relief of tension" (Sullivan 1947: 44).

Experiencing

In Sullivan's system the term experience connotes transformations of energy. Sullivan distinguishes between two types of experience, namely *tension*, which means the aptitude and the readiness to act, and *action* itself, which Sullivan defines as *transformation of energy*.

There are two types of tensions: tensions of *needs* and tensions of *anxiety*. All needs can be divided into general and zonal. The general needs are physiological requirements of man as a living organism—food, water, reproductive activity. The zonal needs develop as a result of interaction between the respective zones of our body and the environment, e.g., the oral zone, the genital, the manual, and the visual. "Zonal needs" correspond vaguely to Freud's genetic zones. Although some of the areas pointed out are identical, there is quite a difference in the role attached to them respectively by Freud and Sullivan. Freud perceived the bodily zones as representative of the phylogenetic evolution of mankind and interpreted their ontogenetic role on the lines of the biogenetic theory. Sullivan perceives the zones as a product of certain life experiences. They are *zones of interaction with the environment* and do not represent phylogenetic developmental stages.

The second kind of tensions are the *anxiety tensions*. As mentioned before, anxiety is first manifested in early infancy.

> Very young infants show grossly identical patterns of behavior when they are subjected to "frightening" situations and when they are in contact with the person who mothers them and *that person* is anxious, angry, or *otherwise disquieted*. Something which develops without a break into the tension state which we have discriminated on the basis of its specific difference from fear can be *induced* in the infant by *interpersonal influence*, in contrast to the evocation of primitive fear by sundry violent influences from "outside" the infant's body. This *interpersonal induction* of anxiety, and the exclusively interpersonal origin of every instance of its manifestations is the unique characteristic of anxiety and of the congeries of more complex tensions in later life to which it contributes. (Sullivan 1948)

The Three Modes

Under the influence of George Herbert Mead, Sullivan suggested three levels or "modes" of experience, namely the *prototaxic* (which corresponds to Mead's), *parataxic* (the term borrowed from Thomas Verner Moore), and *syntaxic*.

The first mode is prototaxic; the earliest experiences in human life are prototaxic. The infant's experiences are undifferentiated, having no division into time units, since he has not yet developed the awareness of being a separate entity. The prototaxic mode is a series of "momentary states," mostly unformulable and therefore incommunicable. Events come and go, and the infant is unable to localize them in time or space. This experience is a sort of mass experience, unorganized, dim, as if made out of one piece. Some such experiences may be mystical and of "cosmic identification." "The one relationship which certainly exists between items of experience in the prototaxic mode is succession, place in organismic or biological time." As the child grows, he learns "that objects which our distance receptors, our eyes and ears, for example, encounter, are of a quite different order of relationship from things which our tactile or our gustatory receptors encounter." The child learns to differentiate between respective parts of reality. *Parataxic* experience is prelogical, since the child is unable to relate things and events to each other or to understand the laws of nature. Things are perceived without logical course or order. Experiences "take on personal meaning" and become organized into "personification of myself." The parataxic mode of experience remains in adults' life in dreams, when things come and go without being logically connected. In childhood most waking experiences are of this kind.

Children's language is "autistic," i.e., centered around "myself." Everything is related to personal needs and experiences. Language is a system of symbolic signs, but children's symbols are a product of their personification of themselves and of attaching imaginary traits to other people; children's symbols are therefore subjective, personal, and often imaginary. Step by step the child learns to associate certain signs with the behavior of other people and to pay attention to their experiences too. Children's parataxic mode and autistic talk give way to the syntaxic mode and interpersonal language.

The syntaxic mode is related to perceiving other people and validating one's experience against the experience of others. One's observations and judgments are "consensually validated" with perceptions and conclusions of others. Sullivan is aware of the fact that even checking one's experiences against those of others cannot serve as evidence of absolute truth. It means only some kind of agreement with significant persons and makes possible communication and adjustment in a given social environment and culture. Of course, syntaxic experiences depend on cultural settings. Tensions may occur in either parataxic or syntaxic modes. Any frustration in seeking to satisfy hunger, or sex, or the need for physical proximity may be experienced in the syntaxic mode, or in the unorganized and highly imaginary parataxic mode. *Anxiety is always perceived in the parataxic mode.*

Interpersonal Relations

Sullivan's concept of interpersonal relations implies dynamic interaction, quite resembling Kurt Lewin's conceptual system. Sullivan's close associate, Patrick Mullahy, explained Sullivan's concept as follows:

> Quite early in his professional life he arrived at his interpersonal reorientation, as the following quotation from an article published in 1933 in the *Encyclopedia of the Social Sciences* makes clear:
>
> This approach [which he worked out] recognizes that the person, psychobiologically conceived, maintains organization, communal existence, and functional activity in and within both the physicochemical and the superorganic cultural universe. The study of the life course of the individual becomes more intelligible when personality is conceived as the

> hypothetical entity which manifests itself in interpersonal relations, the latter including interactions with other people, real or fancied, primarily or mediately integrated into dynamic complexes; and with traditions, customs, inventions, and institutions produced by man. (Sullivan 1962, pp. 301–302)

Mullahy goes on to say that along with the elaboration of physiochemical factors, there is a progressive elaboration and differentiation of motives. The latter are acquired from one's experiences with a steadily expanding series of "culture surrogates," such as the mother, the family group, teachers, companions, chums, friends, love objects, enemies, employers, and colleagues. The motives manifest themselves in the integration of total situations involving two or more people, "real or fancied," and a variety of cultural "elements" or patterns. Within these motives there are said to be demands for certain activities, sometimes consciously formulated in terms of a goal and other times not consciously formulated, so that activity is unnoticed by the participants.

Sullivan's use of the concept of "interpersonal relations" is perhaps not easy to grasp unless one has studied his lectures with care and an open mind. In trying to teach the theories based on this concept, I have often suggested the analogy of a tennis match. Almost everything that one participant does is in relation to what the other one does or is anticipated to do, and conversely. While the analogy is by no means perfect, it suggests the reciprocal interplay of human actions in situations where one is dealing with others (Chrzanowski 1977).

But Sullivan did not confine his concept of interpersonal relations to actual situations where two or more people are involved. It includes situations where only one of the "participants" is physically present. The other may be physically absent, as, for example, when a diffident adolescent girl rehearses in advance what she will or will not say to her date and what she will or will not do or, subsequently, when she reviews in her mind what she should or should not have said or done. Still again the "other" may have no physical existence but be strictly an "imaginary" person, that is, a complex symbolic, personified abstraction of past interpersonal experience. In this connection, novelists and schizophrenics come easily to mind. Finally an individual may perceive another person not as he appears to an objective observer but as a blend of qualities he has with qualities some other person in the individual's past life has had or even with qualities of an imaginary person. In a celebrated article, "Psychiatry: Introduction to the Study of Interpersonal Relations," Sullivan has shown how illusory "me-you" patterns function in interpersonal relations.

Clearly, Sullivan has tried to incorporate "intrapsychic" processes in his theory of interpersonal relations. In other words, following the lead of such men as Charles Horton Cooley and George Herbert Mead, he tried to show how the social world of which one is inextricably a part remains influential even when one is physically alone. This private world may or may not become public. It does become public to the extent that it is communicated.

Personality Theory

Sullivan did not believe that personality could be conceived outside of the social context. "Personality can never be isolated from the complexity of interpersonal relations in which the person lives and has his being." Personality is a produce of interrelationship with other persons; therefore it can hardly be regarded as a separate entity. Moreover, personality per se eludes scientific investigation, since it cannot be perceived outside the interpersonal relationship.

"In the course of psychiatric inquiry one discovers that it is not a person as an *isolated and self-contained* entity that one is studying, or can study, but a situation, an interpersonal situation, composed of two or more people." The traits which characterize the interpersonal situations in which one is integrated describe what one is. "Personality is . . . a function of the kinds of interpersonal situations a person integrates with other, whether real persons or personifications," comments Mullahy on Sullivan's theory of personality.

Sullivan presents personality in two modes: in a "private mode," which is incommunicable, and in which no personality can be studied, and as a "relatively enduring pattern of recurrent interpersonal situations." The latter really deserves the label of personality. This "enduring pattern" is personality as grasped by scientific tools.

The "enduring pattern of behavior" is the way a person uses his energies. Here the conception of *self-dynamism* or *self-system* or *self* enters the picture. Sullivan thinks about energy in physical terms; to him dynamism means a "relatively enduring configuration of energy."

This energy is released in interpersonal relations. Self-dynamism is therefore the only part of personality which is observable and can be scientifically studied (Wittenberg 1973).

The directions of use or release of energy are a product of two basic factors. The first is the child's need for satisfaction and security. The second is the influence of the significant persons on the child's behavior. In other words, the child strives to use his energy in order to obtain the satisfaction of his needs and the feeling of euphoria. His activities are determined by the approval or disapproval of significant people. He avoids their disapproval, which brings him anxiety, and tends to seek satisfaction and security in a way which produces social approval.

Self-Dynamism

One of the key concepts in Sullivan's theory is the self-system of self-dynamism. The self-dynamism is the core of personality. It is the conscious part of personality. According to Sullivan the self-dynamism

> permits a minute focus on those performances of the child which are the cause of approbation and disapprobation, but very much like a microscope, it interferes with noticing the rest of the world. When you are staring through your microscope, it interferes with noticing the rest of the world. When you are staring through your microscope, you don't see very much other than that which comes through that channel. So with self-dynamism. (1947: 23)

Self-dynamism is conscious, while anything else is not. Everything outside self-dynamism is either *disassociated* or *selectively inattended*. The first term corresponds to Freud's unconscious, while the latter resembles the preconscious. Only those processes which are pertinent to relationships with significant people become included in self-dynamism and the person is "wittingly aware" of then. Tension and energy transformation may be a "felt or wittingly noted state of being" or not. Tension per se is potentiality for action; energy transformations are actions; both tensions and energy transformations may have "felt or representational components" or may "transpire without any witting awareness."

Mullahy described the self-dynamism as follows:

> Self-esteem is a function of the self-dynamism, which evolves in great part, according to Sullivan, owing to the necessity of warding off anxiety and, as one grows, of protecting self-esteem. But anxiety is induced by "significant people," such as the parents, who embody in their own attitudes, thoughts, and behavior certain limitations and irrationalities of their society. The self is not synonymous with that to which we refer when we say "I" or "me," namely, the personified self. In addition to the personified self, the self-dynamism includes those processes and activities which we fail to attend or "forget," owing to selective inattention, or which we simply fail to formulate. Briefly the self is the locus of meaningful organization of life experience. In the main, the self is a product of one's experiences in interpersonal relations, especially those pertaining to the pursuits of satisfaction and security. The self does not

include everything that one has lived, undergone, or "prehended." Owing to one's personal history, which is ordinarily closely connected with the social order of which one is a member, he cannot make sense of everything he encounters or "prehends," nor can anyone else. Only, or chiefly, those experiences which have meaning for the person in terms of his life history are incorporated into the self and become material for further enriching or more or less disabling experiences.

And further on:

Thus, once more, the *self* is not synonymous with *personality,* for some experiences, including motivational factors and behavior patterns, occur outside awareness and cannot ordinarily become conscious. These are the dissociated aspects of personality or, in traditional psychoanalytic language, deeply unconscious systems. They cannot be recognized by the self. One cannot usually assimilate them because they do not make sense in terms of one's past experience and one lacks real foresight as to what they portend. Indeed they are often terrifyingly antipathetic to the individual's self and his values. . . .
There still remains the question of how the two aspects of personality, the two "systems," remain in balance. One answer is that normally the self-system has sufficient energy at its disposal to keep the forces of the dissociated system functionally isolated. Still another and more specific answer is that the self-system has at its disposal an instrumentality which one unwittingly employs to limit and restrict awareness, namely, selective inattention. . . . The more one has been exposed to the irrational and un-understandable prescriptions of behavior laid down by parents and others, the more restrictive the self tends to be, and the more prone to anxiety and conflict one is likely to be. Stated in another way, if in the course of one's upbringing one has been chronically subjected to severe anxiety in connection with one's behavior, selective inattention and dissociation will have to be employed more extensively, sometimes to a point where one's ability to achieve the satisfactions and security one needs is gravely undermined. A neurosis or, in some instances, a psychosis ensues. (Mullahy 1965: 366–367)

Period of Growth

The First Stage: Infancy

Sullivan suggested to divide the entire period of growth into five stages, from birth to adulthood. These phases are determined by patterns of mother-child interaction.

The mother's nipple is the first source of food for a hungry infant, and the infant's view of the world and of himself is greatly influenced by his initial encounters with the nipple. If the infant's hunger is met by an affectionate and secure mother, the infant perceives the world as a pleasant and secure place to live. He develops a wholesome view of himself and of his needs and regards himself as efficient and successful.

In Sullivan's own words:

> The good and satisfactory nipple-in-lips which is the signal—the uncomplicated signal—for nursing.
>
> The good but unsatisfactory nipple-in-lips which is a signal for rejection until the need of hunger is great enough to make this good but unsatisfactory nipple acceptable.
>
> The wrong-nipple-in-lips, that is, one that does not give milk any longer—which is a signal for rejection and search for another nipple.
>
> The evil nipple, the nipple of the anxious mother which, so far as the infant is concerned, is a nipple preceded by the aura of extremely disagreeable tension-anxiety, which is a signal for avoidance, often even the avoidance of investing the nipple with the lips at all. So that signal might be converted into rather adult words by saying it is a signal for "not that nipple in my lips." (Sullivan 1953: 80)

According to Sullivan, the infant is capable of perceiving his mother's moods. This ability for nonverbal communication, *empathy*, guides the child's reactions to his mother and greatly influences his views of himself and of the world around him. The infant's self or, as Sullivan calls it, *self-dynamism* develops through approval or disapproval by significant adults, and especially the mother. The infant's feelings of security and self-acceptance are determined by the attitude of others as he perceives or empathizes it. Sullivan described this relationship as follows:

> Since the approbation of the important person is very valuable, since disapprobation denies satisfaction and gives anxiety, the self becomes extremely important. . . . It has a tendency to focus attention in performances with the significant other person which got approbation or disfavor. And that peculiarity, closely connected with anxiety, persists thenceforth through life. It comes about that the self, to which we refer when we say "I," is the only thing which has alertness, which notices what goes on, and needless to say, notices what goes on in its own field. The rest of the personality gets along outside awareness. (1947: 23)

The infant develops three types of personifications of himself. When his behavior is accepted and praised, he develops self-identification of approval: This is the "good-me." Mild anxiety is conducive to "bad-me" personification of self. Unusual feelings of horror or shock lead to the diffuse and never fully conscious states of "not-me."

In infancy the child begins to develop personification of other people too. The feeling of euphoria is conducive to a "good-mother" personification, while anxiety states bring "bad-mother" personifications. Out of the personifications of significant people grow and develop later *eidetic* people, i.e., imaginary persons.

Nursing is an interpersonal experience. It is more than milk and mother's breasts; it is the warmth of mother's body, olfactory sentience and, above all, mother's friendly and affectionate attitude determine the nature of the nursing experience.

As mentioned above, the self system begins with *good-me;* the beginning personification which organize experiences in which satisfactions have been enhanced by rewarding tenderness. Good-me develops into "I." *Bad-me* on the other hand is the beginning personification which organizes experience in which increasing degrees of anxiety are associated with behavior involving the mother in its interpersonal setting. Bad-me is based on the increasing gradient of anxiety. *Not-me* evolves very gradually since it comes from the experience of intense anxiety, organizations of experiences which, when observed, have led to intense forbidding gestures on the part of the mother and induced intense anxiety in the infant; Not-me becomes a part of the very disturbed way of living.

The dynamism of the self-system does not have any particular zones of interaction nor any particular physiological apparatus. Sullivan's concept of self-esteem somehow corresponds to Freud's ego and it is directed to avoiding and/or minimizing the tension of anxiety.

The infant may discover situations in which he is totally powerless even when mother tries to help. These situations may elicit a particular cognitive process of "selective inattention" which plays a significant role in the child's future coping with hardships—simply by becoming unaware of them.

The Second Stage: Childhood

The second stage, childhood, starts with the acquisition of spoken language and growing contact with people other than the child's mother. The main processes of this stage have been defined by Sullivan as "integration of an interpersonal situation," which is "a reciprocal process in which 1) complementary needs are resolved or aggravated, 2) reciprocal patterns of activity are developed or disintegrated, and 3) foresight of satisfaction or rebuff of similar needs is facilitated" (Sullivan 1953: 198).

Childhood is the era of the most pervasive acculturation processes. The child learns toilet habits, eating habits, cleanliness, and most of the habitual ways of behavior regarded as proper by the society and represented by his parents.

The child learns how to restrain his drive for *power*. Many of his activities are frustrated or prohibited by his parents. Some of his unacceptable desires are *sublimated.* The anxiety-provoking tendencies are diverted into socially accepted channels; they are partially satisfied by fusion with socially desirable ways of behavior.

The unsatisfied and undischarged parts of the drive or impulse find outlet in dreams, daydreams, or regressions to the presublimation situation. Regression usually represents unsatisfied parts of a drive but it is also regression in the person's total behavior.

Quite often the child's efforts to get approval meet with parental lack of patience, fatigue, or some other kind of rejecting behavior. This may lead to distortions of the reality of the child; the child who craves parental love and receives nothing but rejection may develop paranoid distortions. He may regard himself as lonely and rejected in a cold and hostile world.

The child may then try to alleviate his anxiety by producing anger. Outbursts of anger undoubtedly bring relief from frustration, as several studies have proved. A frustrated person may feel better after releasing his anger on an innocent scapegoat.

Sullivan introduced at this point a theory of emotions which is at great variance with Freud's theory. Freud regarded love and hatred as innate emotions; fear was to come later, as a result of experience. Horney regarded acceptance and lack of it, anxiety, as the basic emotions. Sullivan saw in anger a substitute for anxiety; most individuals successfully conceal their anxieties even from themselves, and their anger often serves as an outlet and a cover for anxiety.

The Third Stage: Juvenile

The juvenile stage starts with school attendance. The child spends most of his waking time, at least on weekdays, with his peer group. He may be well accepted by his peers and belong to the "in" group or rejected or even boycotted and ostracized. He may become very "popular with the group," and be "one of the boys" or "one of the girls," or may be criticized, ridiculed, and even persecuted.

School-age children can be kind and considerate to one of their peers and intolerant and cruel to another. Thus, the juvenile stage teaches the child to assess the social position and status of himself and of others.

Sullivan summarized the juvenile age experiences as follows:

> To the extent that the juvenile knows, or could be easily led to know what needs motivate his relations with others, and under what circumstances these needs—whether they be for prestige or for anything else—are appropriate and relatively apt to get by without damage to self-respect, to this extent the person has gotten a great deal out of his first big plunge into socialization. (1953: 224)

The Fourth Stage: Preadolescence

Sullivan's preadolescence (age eight-and-a-half to twelve) corresponds to Freud's later part of latency period. This is the time when children form close interpersonal relations in the form of dyads, triads, or larger cliques and groups. As a rule, bonds develop between children of the same sex.

Patrick Mullahy (1965: 362) described preadolescence as a period in which social needs come to the fore. The preadolescent needs tenderness, adult approval of and participation in the child's activities, association with peers, acceptance, and, above all, the need for a close and intimate relationship with someone.

"Because one draws so close to another," Sullivan wrote, "because one is newly capable of seeing oneself through the other's eyes, the preadolescent phase of personality development is especially significant in correcting autistic, fantastic ideas about oneself or others" (1953: 248).

The Fifth Stage: Adolescence

At this phase of development the *lustful* behavior comes to fore. Sullivan introduced a distinction between the capacity for interpersonal relations and intimacy with a person of the opposite sex on one hand, and the development of sexual behavior proper. It is typical for this age not to turn to parents for guidance but to seek information from peers. Adolescent boys first explore the meaning of sex with other boys whom they esteem and regard, and girls turn to their female peers. Even the steps toward intimacy with the opposite sex take place with close friends of the same sex. Girls discuss the boys they are beginning to care for, as well as boys per se, with their girl friends; boys do likewise. If no undue warping has occurred, boys and girls both discover a new kind of intimacy with a very special person of the opposite sex. Romantic love or infatuation may then emerge as the chosen person becomes more important than one's friends. Disappointment and disillusion in romantic love may occur as the intimacy deepens and incompatabilities or other difficulties emerge between the two persons. The pain of disappointment may provoke a fearful withdrawal for a short or longer period of time, in which one's friends will again become a great resource. One learns from a series of such experiences, each deepening one's capacity for love and preparing for the challenge of the next relationship, leading, hopefully, to a mature and enduring love between two persons of the opposite sex (Green 1972: 525).

Mental Disorder

Sullivan perceived the parent-child interaction as the main cause of mental disorder. Then nonloving, punitive, rejecting parents may prevent the development of the self-system in the child and force mass dissociation. When these adverse processes produce a state of panic and the "not-me" feeling, the child is pushed into schizophrenia (Sullivan 1947, 1953).

According to Sullivan, "Personality can never be isolated from the complexity of interpersonal relations in which the person lives and has his being." In psychiatric inquiry we do not study a person as *an isolated and self-contained entity,* but the interpersonal situation. Sullivan, without using the term cathexis, emphasized empathy, described by him as a kind of "emotional contagion or communion" between the child and the parental figures. He wrote, "The infant shows a curious relationship or connection with the significant adult, ordinarily the mother. If the mother . . . is seriously disturbed around the time of feeding, then on that occasion there will be feeding difficulty or the infant will have indigestion" (Sullivan 1947: 7).

Sullivan discovered a great truth when he wrote that schizophrenia is a catastrophe is self-esteem (1947, 1962). The tormenting feelings of one's own worthlessness, guilt, and inadequacy are common to all schizophrenics. On a prepsychotic level schizophrenics doubt their appearance, intelligence, honesty, and blame themselves for whatever hardships they have had. When seriously deteriorated, they do not care about their physical appearance, their jobs, income, food, or health. "When I realized how bad I was, I punished myself by not eating," explained a thirty-five year old schizophrenic in remission. When the superego attacks the ego, the ego may accept some additional, external punishment to alleviate the attack. Schizophrenics often provoke fights, expecting defeat and accepting it as a deserved punishment.

BIBLIOGRAPHY

Chrzanowski, G. 1977. *Interpersonal Approach to Psychoanalysis*. New York: Gardner Press.

Green, M. R. 1972. The interpersonal approach to child therapy. In B. B. Wolman, ed., *Handbook of Child Psychoanalysis*, pp. 514–568. New York: Van Nostrand-Reinhold.

Lewin, K. 1935. *A Dynamic Theory of Personality*. New York: McGraw-Hill.

Mullahy, P. 1948. *Oedipus, Myth, and Complex*. New York: Hermitage Press.

—— 1965. Non-Freudian analytic theories. In B. B. Wolman, ed., *Handbook of Clinical Psychology*, pp. 341–380. New York: McGraw Hill.

Spiegel, R. 1977. Sullivan, H. S.—Historical review. In B. B. Wolman, ed., *International Encyclopedia of Psychiatry, Psychology, Psychoanalysis, and Neurology*, 11: 26–31. New York: Aesculapius.

Sullivan, H. S. 1947. *Conceptions of Modern Psychiatry*. Washington: White.

—— 1948a. The meaning of anxiety in psychiatry and in life. *Psychiatry*, 11: 1–13.

—— 1948b. Towards a psychiatry of peoples. *Psychiatry*, 11: 105–116.

—— 1953. *The Interpersonal Theory of Psychiatry*. New York: Norton.

—— 1954. *The Psychiatric Interview*. New York: Norton.

—— 1956. *Clinical Studies in Psychiatry*. New York: Norton.

—— 1962. *Schizophrenia As a Human Process*. New York: Norton.

Wittenberg, E. G. and L. Caligor. 1967. The interpersonal approach to treatment. In B. B. Wolman, ed., *Psychoanalytic Techniques*, pp. 424–441. New York: Basic Books.

Wittenberg, E. G., ed. 1973. *Interpersonal Explorations in Psychoanalysis*. New York: Basic Books.

[EIGHTEEN]
Quo Vadis, Psychoanalysis?

"THE way in which great writers and poets deal with human life presents psychology with a task and a material. Here is the intuitive understanding of the whole system which psychology has to approach in its own way, by generalization and abstraction. One wishes for a psychology which could catch in the net of its descriptions that which these poets and writers contain over and above present-day psychology; a psychology which takes the thoughts which Augustine, Pascal, or Lichtenberg make so penetrating by one-sided brilliant illumination, and make them serviceable for human knowledge in a universally valid system," wrote Wilhelm Dilthey in 1894 in the *Ideen über eine beschreibende und zergliedernde Psychologie.*

The man who developed such a system was Sigmund Freud. After Freud, no one could complain that *King Lear, Hamlet,* and *Macbeth* contained more psychology than psychoanalysis. For right or wrong, Freud did answer the question: What are the forces that make men and women act sanely or insanely, like angels or animals, saints or sinners, heroes or cowards? Freud's penetrating eye did not stop at the study of overt behavior; Freud unraveled the unconscious motives of human actions and discovered what makes human beings think, feel, desire, and act.

After a long and relentless search Freud arrived at the idea of two driving forces: the force of life, Eros, and the force of death, Thanatos. Men are born to live and to die, and these two forces merge, fuse, and clash in a tragic conflict of life and death.

Freud was a pioneer and an iconoclast. Not everything he said is immortal, but the fact that he said it is immortal, for the courage to say what one thinks and the never-ending search for truth is immortal.

Freud was a lonely researcher in the *Spiessbürger* old Vienna, in that ultra-Victorian, prudish, bigoted society. One man against the establishment. One human mind against the official science, one thinker against an ocean of prejudice, superstition, and pompous ignorance. Freud opened his heart to show others how human hearts beat. He laid bare his own weaknesses and vicissitudes so that others could understand their weaknesses and their vicissitudes. He tore up the web of lies sanctioned by tradition and ignorance.

In 1939 Freud passed away. In his last years he wrote his most controversial works, *Civilization and Its Discontents* (1929), *Future of an Illusion* (1927), *Moses and Monotheism* (1934), and *Why War* (1932). He did not spare the holy cows of any creed and did not share the naive optimism of pacifists. All his life he was an uncompromising, nonconformist, independent thinker. Intellectual freedom and moral courage were his way of life. This was his historical mission.

Psychoanalysis has since become a comfortable profession. Medical psychoanalysts are today well organized and respected by the medical profession. Nonmedical analysts successfully fight for and are gradually winning approval and respect through their proficiency in practice and research. Freud's *Interpretation of Dreams* was printed in 600 copies and it took eight years to sell them. Today leading publishers are delighted to publish psychoanalysis books. There are scores of psychoanalytic journals. Psychoanalysis is taught in practically all leading universities and has become an integral part of graduate and postdoctoral medical, psychological, and social work programs.

Where should it go from now on? Should it repeat the great teachings of the Master, adding talmudic comments to the scripture, and following religiously the letter of his great books?

Quo vadis, psychoanalysis?

Jesus walked on the hills of Judea. He went to the people. But his disciples created the organized power, the Ecclesia.

Zarathustra left the desert and came to the town to teach people. But his disciples developed the Manicheistic cult.

Must psychoanalysts "follow" Freud or should they start where Freud stopped?

Radicals of yesterday can easily become today's conservatives. The fact that Freud's works created a revolution at the beginning of our century, through his *Interpretation of Dreams, Three Essays on Sexuality*, and *Totem and Tabu,* does not make his disciples today into revolutionaries. Early Christians were revolutionaries, but were no longer so in times of Giordano Bruno and Galileo.

History has no mercy for epigones nor respect for fossils. Progress is not a one-time achievement, but a continuous process of craving and striving. There are no final achievements in the history of the human mind.

Some of the efforts to go beyond Freud's teachings are described in this volume. Are they closer to truth than Freud's system?

No one has the monopoly on truth, but obviously the theory Freud built requires a fresh look. Do we really believe in the Kraepelinian classificatory system? Do we have adequate evidence for hooking etiology to the phenomena of fixations following Karl Abraham's developmental phases?

J. Arlow, L. Brenner, M. Gill, and others critically examined Freud's topographic theory. R. R. Holt, K. Pribram, and others analyzed Freud's biological and neurological concepts. Today we are in possession of empirical data that were unavailable in Freud's time. Freud never hesitated to modify his theory whenever new empirical data suggested that such a modification might be advisable. Freud never bent facts to fit into a theory; he was never an orthodox Freudian. Must we be?

All his life Freud was keenly aware of social, cultural, and moral issues. He crossed words and swords with Judaism, Christianity, Marxism, surrealism, and communism. He could, indeed, say with Terentius, "Homo sum, humani nil a me alienum esse puto" (I am a human and I believe nothing human is alien to me). Freud was deeply involved with problems of psychology, anthropology, history, sociology, biology, theology, literature, and art.

Some scientists shy away from the irrationality of human behavior. Some fear the unconscious. Some prefer to hover over the surface of the volcano of human feelings instead of digging deep in search for truth.

But nothing could stop Freud. His mission was the mission of relentless truth. Freud's heritage is to face the social and cultural issues of our times. In our times the issues are not sexual prudishness. Our problems are problems of the atomic age and of space exploration.

Freud developed his ideas in opposition to Victorian times. We live in times of rapid expansion of communication, liberation of developing countries, civil rights struggles, urban development, and dehumanizing technology. It is not enough to study Freud's analysis of *his* times. A psychoanalytic analysis of our times, of our art, literature, religion, and education is imperative.

Our times are times of high technological development and rapid change in cultural and moral values. Freud's generation grew in an atmosphere of too much restraint and too many rigid inhibitions. Freud struggled against the lies of conformity of his times; psychoanalysts today face a new pseudosophisticated conformity to an alleged nonconformism.

Psychoanalysts are not social reformers. They must not form a church or a political party. Freud doubted whether any particular social system can solve all the ills of human nature. Freud did not subscribe to a Tolstoi, or a Marx, or a Kropotkin, or any ideal or perfect society.

Prometheus went to Olympus and stole the gods' fire. At that time fire was the privilege of the gods. Prometheus stole it and gave it to men who lived in cold and darkness. He was punished by Zeus. He was fastened to a rock and an eagle cut his chest open and ate his liver.

Freud opened his chest and laid his heart bare. He gave himself to humanity. He was never a moralist; he was just as moral as a scientist should be.

Freud's teachings might be wrong on many issues, but his "Sapere aude!" (Have the courage to know) is right forever.

NAME INDEX

Abraham, K., 118, 120, 131, 134, 325
Abrams, S., 64
Adler, A., 4, 23, 31, 50, 110, 207, 209, 234, 251-65, 267, 268, 303
Adler, K., 256, 257, 261
Ajdukiewicz, K., 10
Allport, G. W., 32
Amsel, A., 62
Ansbacher, H. L., 252, 254
Ansbacher, R., 252, 254
Antigone, 245
Aquinas, T., 61
Archimedes, 3, 13
Arlow, J., 65, 85, 325
Augustine, Saint, 323
Ayer, A. J., 10
Ayrapentyantz, E., 30

Bacon, F., 69
Barker, R. G., 15
Beard, G. M., 46
Bell, C., 43
Benecke, F. E., 52
Bergmann, H., 24
Bergson, H., 53
Berkeley, G., 8, 12
Bernard, C., 43
Bernfeld, S., 45
Bernheim, H. M., 78
Bettelheim, B., 239
Binet, A., 29
Blanck, G., 152, 175, 217
Blanck, R., 174, 217
Bowlby, J., 219
Boyle, R., 4, 13
Braid, J., 78
Braithwaite, R. B., 10, 24
Brenner, C., 65, 85, 325
Brentano, F., 252, 254
Brett, G. S., 9
Brücke, E., 44, 46, 68
Bruno, G., 324
Burt, C., 292
Bykov, K. M., 30

Cannon, W. B., 229, 310
Carnap, R., 10, 11
Cassirer, E., 7, 24
Charcot, J. M., 46, 78
Child, I. L., 240, 243
Chrzanowski, G., 313
Cleopatra, 280
Cohen, M. R., 13, 26
Coleridge, S. T., 51
Columbus, C., 12
Comte, A., 8, 9, 12, 21
Cooley, C. H., 314
Copernicus, N., 3, 12
Cranefield, P. F., 44

Darwin, C., 3, 4, 27, 29, 31, 41, 42, 49, 68, 79, 230, 232, 252
Daudet, A., 29
Dembo, T., 15
Dement, W. C., 18
Descartes, R., 4, 5, 6, 25, 51, 61, 65
Dewey, J., 65
Dilthey, W., 49, 50, 251, 323
Dingle, W., 19, 24
Dollard, J., 15, 232
Dostoyevsky, F., 49
Du Bois-Reymond, E. H., 44

Duhem, P., 24
Dunlap, K., 40
Durkheim, E., 275

Eaton, R. M., 61
Ebbinghaus, H., 13
Edison, T., 40, 49
Einstein, A., 4, 20, 21, 22, 24, 25, 68, 69, 229
Ellenberger, H. F., 53
Erikson, E. H., 206-13, 232, 243
Evans, R., 292

Fechner, G. T., 13, 44, 45, 73
Federn, E., 110
Feigl, H., 10, 14, 32
Fenichel, O., 118, 231, 235
Feyerabend, P. K., 3
Flourens, P., 43
Fox Bourne, H. R., 5
Frank, P., 10
Freud, A., 177
Freud, S., 3, 4, 11, 13, 17, 18, 19, 20, 21, 22, 29, 30, 31, 32, 33, 34, 40, 41, 44, 45, 46, 47, 48, 49, 50, 51, 52, 53, 61-172, 173, 174, 176, 188, 192, 193, 206, 207, 209, 214, 216, 222, 228, 229, 232, 233, 234, 235, 236, 238, 239, 240, 243, 245, 251, 252, 253, 258, 259, 260, 261, 262, 266, 267, 268, 281, 299, 300, 301, 302, 303, 310, 318, 319, 323, 324, 325, 326
Fromm, E., 31
Furer, M., 215, 225

Galileo, G., 3, 5, 324
Gedo, J. E., 104, 107
Gesell, A., 240, 243
Gill, M., 67, 85, 325
Goedel, K., 10
Goethe, J. W., 51
Goldstein, K., 229
Green, M. R., 321
Griesinger, W., 45, 46
Guthrie, E. R., 15

Haeckel, E., 41, 102, 252
Halverson, H. M., 15
Hamlet, 61
Hartmann, E. von, 53
Hartmann, H., 173-191, 192, 209, 214, 232, 240
Harvey, W., 4
Hegel, G. W. F., 7, 61, 62
Heisenberg, W., 20
Helmholtz, H. von, 27, 44, 45
Hempel, C., 10
Henry, C. W., 78
Herbart, J. F., 23, 45, 49, 51, 83-84, 252
Hertz, H., 40
Hippocrates, 262
Hodges, H. A., 50, 253, 254
Holt, R. R., 85, 145, 325
Horney, K., 33, 238, 239, 299-308, 309
Horton, G. P., 15
Hull, C. L., 20, 33
Hume, D., 5, 6, 10, 11, 19, 24, 25, 26
Humphrey, G., 46, 47
Husserl, E., 7
Huxley, T. H., 41, 49, 68

Ibsen, H., 49
Isaacs, S., 15, 200, 201

Jacobi, J., 282
James, W., 12
Jeffreys, H., 8
Jesus Christ, 276, 293, 324
Jones, E., 51, 162, 163
Jung, C. G., 4, 13, 31, 53, 207, 266-98

Kaganov, V. M., 27
Kanner, L., 224
Kant, I., 6, 7, 8, 9, 10, 11, 21, 25, 51, 52, 61, 65, 69, 168
Kardiner, A., 232, 240
Kelman, H., 300, 302, 306-7
Kerenyi, C., 275
Klein, M., 192-205, 258
Kleist, E. von, 41
Koch, S., 15
Kotarbinski, T., 10
Kris, E., 175, 177, 183, 184, 240
Krishaber, M., 46
Kropotkin, P., 326
Külpe, O., 62

Lamarck, J. B. de, 41, 42
Laplace, P. S., 24
Lawson, R., 232

Leibniz, G., 51
Lévy-Bruhl, L., 275
Lewin, K., 13, 15, 20, 254, 309, 312
Lichtenberg, G. C., 323
Liebault, A. A., 78
Liszt, F., 254
Locke, J., 5, 6, 11, 25
Loewenstein, R. M., 175, 177, 183, 184, 186, 240
Loewenthal, H. S., 110
Lombroso, C., 78
London, H., 235
Lorentz, H. A., 40
Ludwig, C. F. W., 44
Lukasiewicz, J., 10

MacCorquodale, K., 32
McDougall, W., 23
Mach, E., 8, 9, 11, 12, 25, 69
Maeterlinck, M., 49
Magendie, F., 43
Mahler, M. S., 214-27
Maimonides, M., 61
Malthus, R., 42
Mann, T., 53
Marconi, G., 41, 49
Marx, K., 326
Mason, A. S., 17
Maupassant, G. de, 49
Maxwell, J. C., 27, 40
Mead, G. H., 309, 311, 314
Meehl, P. E., 32
Meissner, W. W., 67
Mendel, G., 4, 13, 32, 41, 49
Mendeleev, D. I., 41
Mesmer, A., 41, 77
Meyer, A., 309
Meynert, T., 46
Michelson, A. A., 20, 40
Mill, J. S., 14, 24, 69
Miller, J., 27
Money, J., 126, 233
Moore, T. V., 311
Morgan, L., 27
Morgenbesser, S., 27
Morley, E. W., 20, 40
Moses, 276
Mullahy, P., 312, 313, 315, 316, 320
Müller, J., 43, 44
Muschenbroek, Pieter van, 41

Nagel, E., 13, 14, 19, 22, 23, 67, 70
Nageli, Karl von, 41
Napoleon Bonaparte, 253
Narcissus, 107
Neurath, O., 10
Newton, I., 3, 4, 79
Nietzsche, F., 49, 53
Nunberg, H., 110

Oedipus, 121, 122
Oersted, H. S., 40

Pascal, C. R., 323
Pasteur, L., 41, 49, 79
Pavlov, I. P., 3, 4, 13, 14, 47, 49, 68, 82, 229, 230, 231, 232, 236
Piaget, J., 3, 65, 240, 243
Planck, M., 25
Platner, E., 51
Plato, 74
Platonov, G. V., 27
Poincaré, H., 19, 21, 69
Pollock, G. H., 104, 107
Pribram, K., 325
Progoff, I., 284, 285
Prometheus, 245, 326

Rank, O., 50, 147
Rapaport, D., 67, 174
Reichenbach, H., 10
Rickert, H., 253
Ritchie, B. F., 33
Roosevelt, F. D., 145
Rosenzweig, S., 17, 18
Ross, N., 64
Royce, J. R., 27
Russell, B., 10, 21

Sarris, V., 27
Sartre, J. P., 208
Schelling, F. W. J., 51
Schiller, J. C. F., 51
Schlick, M., 10
Schopenhauer, A., 7, 11, 49, 52, 53
Sears, R. R., 17
Sechenov, I. M., 30, 68
Segal, H., 195, 201, 202
Skinner, B. F., 29, 62, 63
Smuts, J. C., 251
Sophocles, 121

Spencer, H., 9, 41
Spinoza, B., 5, 25
Spranger, E., 50, 254, 261
Stebbing, S., 10
Stern, W., 4, 50, 254
Strindberg, A., 49
Sullivan, H. S., 31, 32, 33, 239, 303, 309-22
Sulloway, F J., 77, 137

Teitelbaum-Tarski, P., 10
Thorndike, E. L., 3, 15, 27, 62
Titchener, E. B., 29, 48
Tolman, E. C., 32, 62
Tolstoi, L., 326

Vaihinger, H., 4, 31, 50, 251, 254, 255
Vollmerhausen, J. W., 300, 302, 306, 307
Vries, H. de, 41

Wagner, R., 254
Wallace, A. R., 41, 42, 49
Watson, J. B., 29, 32, 47, 50
Weber, E. H., 13, 27, 44
Weismann, A., 41, 42
Werthermer, M., 254
Whiting, J. W., 240, 243
Whyte, L. L., 51, 52
Windelband, W., 252, 253, 254
Wittenberg, E. G., 314
Wittgenstein, L., 10, 11
Wolff, C. von, 51
Wolman, B. B., 5, 7, 13, 15, 19, 26, 27, 47, 51, 52, 62, 65, 67, 85, 126, 174, 228-47
Woodworth, R. S., 32
Wordsworth, W., 51
Wotan, 275
Wundt, W., 8, 9, 28, 29, 46, 47, 48, 49, 50, 62

Yerkes, R. M., 29

Zaratustra, 324
Zeigarnik, B., 13
Zeus, 74, 326
Zilboorg, G., 78
Zola, E., 49

SUBJECT INDEX

Adaptation (Hartmann), 179-183
Aggression, 109-13, 183, 184, 194-99, 228-32, 242-44, 257-59; Adler on, 257-59; Freud on, 109-13; Hartmann on, 183-84; Klein on, 194-99; Wolman, on, 228-32, 242-44
Anxiety, 145-49, 173, 192-96, 301, 310-12; Freud on, 145-49, 173; Horney on, 301; Klein on, 192-96; Sullivan on, 310-12
Archetypes (Jung), 275-80
Ares (Wolman), 229-32
Autonomy (Hartmann), 174, 178

Cathexis, 103, 112, 129, 137, 143-44, 236-39; anticathexes, 143-44; Freud's concept of, 103, 112, 129, 137; Wolman's concept of, 236-39
Causality, causal principle, 24-27, 71, 72, 267; critique of, 25-27; definition of (Wolman), 25-26; Jung on, 267; in philosophy, 24-25; in physics, 24-27; in psychoanalysis, 71-72; in psychology, 26-27
Child development, 115-27, 193-200, 209-21, 239-42, 259, 260, 303, 311-12; Adler's theory, 259-60; Erikson's theory, 209-12; Freud's theory, 115-27; Horney's theory, 303; Klein's theory, 193-200; Mahler's theory, 214-21; Sullivan's theory, 311-12, 316-21; Wolman's theory, 239-42
Chinese philosophy, 294-96

Death wish, 110-13, 192-93; Freud on, 110-13; Klein on, 192-93

Defense mechanisms, 149-51, 179-83, 194-202; Freud's theory of, 149-51; Hartmann's theory of, 179-83; Klein's theory of, 194-202
Dreams, interpretation of dreams, 85-96, 260-61, 269-71; Adler's theory of, 260-61; Freud's theory of, 85-96; Jung's theory of, 269-71

Ego, 138-52, 173-89, 192-99, 207-9; Erikson's concept of, 207-9; Freud's concept of, 138-52; Hartmann's concept of, 173-89; Klein's concept of, 192-99
Electra complex (Freud), 123
Entropy (Jung), 268-69
Epigenetic principle (Erikson), 206-7
Eros, 107, 113, 118, 228; Freud's concept, of, 107, 113, 118; Wolman's concept of, 228
Experimental research, 17-18, 27, 29, 30-32

Female psychology, 123-25, 234-35; Freud's concept of, 125-25; Wolman's concept of, 234-35
Future of psychoanalysis (Wolman), 323-26

History of science, 41-50, 61-63; biology, 41-43; neurophysiology, 43; physics, 40-41; psychiatry, 45-46; psychology, 46-50, 61-63

Id (Freud), 137-38
Inferiority feeling (Adler), 255-57

Libido, 103-13, 128-29, 267-69; Freud's concept of, 103, 113, 128-29; Jung's concept of, 267-69
Lust for life, 230-31; Pavlov's concept of, 230-31; Wolman's concept of, 230-31

Morality, moral principles, 168-69, 188-89, 212, 244-45, 312-13; Erikson, 212; Freud, 168-69; Hartmann, 188-89; Sullivan, 312-13; Wolman, 244-45
Motivation, 63-64, 207-8, 300-1; Erikson's theory of, 207-8; Freud's theory of, 63-64; Horney's theory of, 300-1; Wolman's theory of, 230

Narcissism (Freud), 107-15
Nirvana principle (Freud), 109

Oedipus complex, 121-33, 199-200; Freud's theory of, 121-33; Klein's theory of, 199-200

Personality theory of, 136-59, 194-99, 207-12, 235-39, 261-63, 273-88, 302-5, 314-18; Adler's, 261-63; Erikson's, 207-12; Freud's, 136-59; Hartmann's, 174-83; Horney's, 302-5; Jung's, 273-88; Klein's, 194-99; Sullivan's, 314-18; Wolman's, 235-39
Philosophy of science, 5-35, 253-55; as-if philosophy, 254-55; empiricism, 5-6; epistomology, 11-13, 21; evidence, 16-18; explanation, 18-20; Kantianism, 6-7; Kulturwissenschaft, 253-54; logical positivism, 10; operationism, 33-34; phenomenology, 6-10; prediction, 18-20; rationalism, 5; theory formation, 18-23, 32-35; truth, concept of, 19
Power and acceptance (Wolman), 232
Psychopathology, 97-98, 118, 221-25, 263-64, 291-92, 303-6, 321; Adler's system, 263-64; Freud's system, 97-98, 118; Horney's system, 303-6; Jung's system, 291-92; Mahler's system, 221-25; Sullivan's system, 321

Quest for identity (Erikson), 206-9

Religion, 162-64, 290-95; Freud on, 162-64; Jung on, 290-95

Sexuality, 102-6, 123, 232, 294; Freud's views, 102-6, 123; Horney's view, 294; Wolman's views, 232
Social relations, 165-69, 185-89, 235-37, 257-59, 288-90, 294, 295, 312-14; Adler's theory of, 257-59; Freud's theory of, 165-69; Hartmann's theory of, 185-89; Jung's theory of, 185-89; Sullivan's theory of, 312-14; Wolman's theory of, 235-37
Superego, 152-58, 188-89, 195-96, 245; Freud's concept of, 152-58; Hartmann's concept of, 188-89; Klein's concept of, 195-96; Wolman's concept of, 245
Symbolic self (Kelman), 306-7

Thanatos (Freud), 108, 110-13, 118
Theory formation in psychoanalysis, 61-76, 83-84, 109, 138-45; case study method, 66-67; constancy principle, 73-74, 101, 112; determinism, 71-72; economy principle, 72-73; energetism, 70-71; epistemological realism, 68-70; motivation, 63-64; monism, 70; Nirvana principle, 109; outside influences: cultural climate, Herbart, 49, 83-84; pleasure and reality principles, 138-40, 142-45; pleasure-unpleasure, 73-77; rationalism, 61, 65-66; thinking processes, 65-65

Unconscious, 30, 51-54, 79-99, 200, 202, 260-61; Adler's concept of, 269-61; Freud's concept of, 30, 51-54, 79-99; Klein's concept of, 200-2